Alexandre Queiroz Pereira

Coastal Resorts and Urbanization in Northeast Brazil

 Springer

Alexandre Queiroz Pereira
Universidade Federal do Ceará
Fortaleza, Ceará, Brazil

ISSN 2366-763X   ISSN 2366-7648   (electronic)
SpringerBriefs in Latin American Studies
ISBN 978-3-030-46592-6   ISBN 978-3-030-46593-3   (eBook)
https://doi.org/10.1007/978-3-030-46593-3

This Springer imprint is published by the registered company Springer Nature Switzerland AG
The registered company address is: Gewerbestrasse 11, 6330 Cham, Switzerland

SpringerBriefs in Latin American Studies

Series Editors

Jorge Rabassa, Lab Geomorfología y Cuaternario, CADIC-CONICET, Ushuaia, Tierra del Fuego, Argentina

Eustógio Wanderley Correia Dantas, Departamento de Geografia, Centro de Ciências, Universidade Federal do Ceará, Fortaleza, Ceará, Brazil

Andrew Sluyter, Conference of Latin Americanist Geographers, Louisiana State University, Baton Rouge, LA, USA

More information about this series at http://www.springer.com/series/14332

Introduction

The spatial mobility of visitors due to leisure and the construction of specific real estate are explicit vectors of the transformation of tourist locations, especially from the end of the twentieth century and beginning of the twenty-first century. Likewise, the scientific production on this topic is, theoretically and methodologically, vast and varied (Duhamel 2018). This is true for different natural environments, especially coastal areas, where seaside resorts have emerged.

From the twentieth century, most studies have used the term second residence. The term appears similarly in various languages: in Spanish, *residencia secundaria*; in French, *résidence secondaire, maison de vacances*, or *résidence de loisir*; and in English, recreational home and holiday home. Hall and Müller (2004) prudently remind us of the extensive discussion of the terminology and the different names used. The authors edited the 2004 book Tourism, Mobility and Second Homes, which covered different approaches and countries (Australia, Spain, USA, South Africa, Norway, and New Zealand, among others), and is one of the greatest English-language contributions to the field. However, the Latin American (Global South) context was not included and more specifically, neither was the Brazilian scenario. In the last 30 years, public and market investments have transformed the Brazilian coast into a leisure space. For instance, demographic censuses have demonstrated a rise in the number of second homes, and tourism records indicate an increase in the number of developments and beds. In Brazil, domestic tourism, involving mobility between regions, is one of the industry's main features.

This book aims to contribute to this discussion, highlighting the coastline of Northeast Brazil and its inclusion in the world circuit of (re)produced spaces for the different leisure practices on the seashore. The theoretical structure prioritizes the use of the concepts of maritime *vilegiatura* and tourism in association with the analysis of urbanization and metropolization processes. Accordingly, attention is also given to other well-used concepts of the international bibliography (second residences, residential tourism, and *vacancier*, among others) to foster an understanding based on urban and leisure geography.

The meaning of the expression second residence itself is not fully understood and has been the subject of much debate. Boyer (2008) argues that the adjective secondary does not explain the complexity of uses of this type of property. For Pino (2003), the *secundariedad* (secondary nature) of the residence/property is not merely quantitative; it is very much a qualitative subject. Cavaco (2003) names properties second residences, thus adding the notion of dwelling, which in turn refers to "frequenting, presence, permanence, and also rest, with some rights over the property and feeling this" (p. 49).

In the etymological structure of the term *vilegiatura*, the villa or residence is a key component to the origins of the practice. If in the classic Renaissance model of *vilegiatura* the construction of the property was a sine qua non-condition, in modern times this is no longer the case. There is a multiplicity of new standards and developments that condition flexibilization, giving the individual the option of renting or sharing the property(ies). This malleable situation brings together both less-affluent family groups, who rent a house for a weekend on a beach near their usual place of residence (short displacements, approximately 100 km), and foreigners, who stay in resort-type ventures thousands of miles away from their homes. Aledo (2008) uses the term residential tourism for the latter situation.

Nowadays, there are a variety of ventures, such as real estate-tourist complexes and hotels (resort, residential condominiums, hotels, clubs, parks, golf courses) that serve a varied clientele. Enterprises of this magnitude have expanded the services offered, restructuring their internal spaces, bringing together tourists and vilegiaturists, renters, and owners. Also, some spaces reflect the linguistic pattern of *vilegiatura*, traditional "do-it-yourself," minimalist second homes.

Directly related to travel, *vilegiatura* defines a non-existential, seasonal, discontinuous, and recreational condition. The individual or group that experiences it aims for enjoyment through their stay, their leisure time, and the activities that suit their tastes. In short, in modern times, *vilegiatura* is characterized by the distinct ways in which social groups organize the practice of a temporary stay for leisure purposes. In this context, the production of second residences is the apex of what Marc Boyer calls *sédéntarité* (making the most of the stay, close to the state of residing). Associated with the relevance and the intensity of tourist mobility, the study of the temporary stay contributes to the reflection on social, urban, real estate, environmental, and cultural transformations in the places currently visited by tourists.

Leisure, in general, and tourism and *vilegiatura*, specifically, are critical causes of the urbanization process. This simple finding leads to a broader analysis that goes beyond counting residences for occasional/seasonal use or even using other notions, such as residential tourism. The need to own and/or use this type of property is a manifestation of how society perceives the day-to-day, time, space, nature, and sociability, thus reproducing peri-urban and metropolitan spaces.

In theoretical-methodological terms, the debate on leisure leads to a discussion of time-space as parts of a social unit, an urban way of life. Dialectically, the way of life conditions and brings together how social segments conceive of and define their space-time and the totality of their social practices, including leisure. In this way, in maritime spaces, *vilegiatura* is interpreted as a practice that begins in urban areas, either as the place of origin of the *vacanciers* (city dwellers) or by the mainly metropolitan urban fabric produced in the destination areas of tourism flows.

Like the city, *vilegiatura* and the use of second residences for leisure are pre-modern and pre-capitalist social products. At present, *vilegiatura*, tourism, and leisure are undeniably overlapping and strategic practices for the (re)production of spaces and spas, especially in the Mediterranean and tropical coastal metropolises.

Against this background, this book intends to highlight the current developments in these socio-spatial practices on the metropolitan coast of the Northeast of Brazil, highlighting the main urban, spa, and tourist agglomerations: Salvador-BA, Recife-PE, Fortaleza-CE, and Natal-RN. Without a doubt, this will lead to an understanding of the processes of urbanization and metropolization associated with maritime leisure.

The following four chapters are the product of approximately 10 years of research carried out in conjunction with colleagues and students of the Laboratory of Urban and Regional Planning of the Federal University of Ceará, Graduate Program in Geography and the Metropolises Observatory.

In the first chapter, the reader will find a historical and conceptual exposition highlighting the relevance of leisure practices, their forms-flows, and their role in the formation of maritime resorts. For this, the concept of *vilegiatura* is used together with the notion of tourism. *Vilegiatura* is a mass socio-spatial practice, which is interwoven with other important concepts such as second residences and residential tourism. The result is an intensified study of urban geography and leisure, aiming to strengthen the understanding of the formation of metropolitan coastal resorts worldwide and, more specifically, in the Northeast of Brazil.

The second chapter analyzes the context of the northeastern region of Brazil and demonstrates the process of modernization and formation of the seaside function within the cities, and later, in the maritime metropolises of the region. The transformations between the start of the twentieth century, the second half of the twentieth century, and the dawn of the twenty-first century are evident. The occupation of the northeastern coastal areas is directly influenced by the global phenomenon of the increasing value and valuation of the urban and metropolitan seafronts of the tropical coasts. This chapter describes the analysis of the genetic components of the process of coastal urbanization in the Northeast, focusing on the socio-spatial variants responsible for the organization of the seafronts of the cities of Salvador, Recife, Fortaleza, and Natal, as well as the metropolitan coasts associated with them. These four urban agglomerations are incontestable and prominent examples of the relationship between tourism and urbanization.

The relationship between metropolization and touristic-real estate ventures is the central theme of the third chapter, which proposes a specific methodology for studies of this nature. It starts from the empirical-conceptual construction of the process and establishes two dimensions of analysis: urban morphology and social practices. Tastes and, consequently, performing maritime practices produce diverse forms and functions, resulting in spatialities marked by leisure and the fragmentation of the urban fabric. In the first decade of the twenty-first century, these socio-spatial transformations have become an essential urban feature in the definition of the northeastern coast. These reflections deal with the region, without forgetting to establish connections with other spaces in Brazil and overseas.

The final chapter presents the seaside resorts in the metropolitan area of Fortaleza, a case study similar to others in the Northeast, examining the urbanistic effects and the key ideas of the planners. The management and social effects resulting from this urbanization model are also investigated, with an emphasis on the ensuing problems. Using quantitative-qualitative research techniques, the analysis on the intra-metropolitan scale explores the developments of coastal urbanization in locations in the metropolis of Ceará. However, these results can be generalized at the regional level, since there is a social and temporal synchronism with the other coastal urban centers in the Northeast (Natal, Recife, and Salvador).

We are grateful for the financial support granted by the National Council for Scientific and Technological Development (CNPq), through the Universal Edict. Likewise, we are grateful for the resources from the Coordination of Improvement of Higher Education Personnel (CAPES), through the Program of Support to Post-graduate and Scientific and Technological Research in Socioeconomic Development in Brazil (PGPSE, n. 88887.123947/2016-00). To the Coastal Environmental Systems and Economic Occupation of the Northeast Project and CAPES PRINT (88887.312019/2018-00), project Social and environmental technologies and methodologies integrated with territorial sustainability—community alternatives to climate change, both coordinated by Professor Jehovah Meireles.

Lastly, I would like to extend my gratitude to my friends and supporters, Eustogio Dantas, Clelia Lustosa Costa, and José Borzacchiello da Silva. All three are examples of dedication to geography and, above all, generosity to those around them.

Welcome to all readers!

References

Aledo A (2008) De la tierra al suelo: la transformación del paisaje y el nuevo turismo residencial, Arbor Ciencia, Pensamiento y Cultura, CLXXXIV, 729, enero-febrero, pp 99–113
Boyer M (2008) Les villegiatures du XVIe au XXIe siècle: panorame du tourisme sédentaire. Paris: Éditions ems
Cavaco C (2003) "Habitantes" dos espaços rurais. Revista da Faculdade de Letras—Geografia. I série, vol. XIX, pp 47–64
Duhamel P (2018) Géographie du Tourisme et des Loisirs. Paris: Armand Colin
Hall C, Müller DK (2004) Introduction: second homes, curse or blessing? Revisited. In: Hall C, Müller DK (orgs.). Tourism, mobility and second homes: between elite landscape and common ground. Clevedon (UK): Channed View Publications, pp 3–14
Pino J (2003) Aproximación sociológica a la vivienda secundaria litoral. Scripta Nova. Universidad de Barcelona. Vol. VII. 146 (026)

Contents

Chapter 1
The Urban, Spas, and Maritime *Vilegiatura*

Abstract The term *vilegiatura* is Italian in origin, villeggiatura, and dates from the Renaissance period. In France, it is commonly used in scientific literature, *villégiature*. In the following introduction, *vilegiatura* is a mass socio-spatial practice existing in complementarity with tourist trips. Other important concepts such as second residences and residential tourism are interwoven with this notion, in order to deepen the studies of urban geography and leisure, chiefly the formation of metropolitan seaside resorts in the world and, more specifically, in the Northeast of Brazil.

Keywords Tourism · Recreation · Resorts

1.1 *Vilegiatura* and Its Urban Nature

To begin with, what is *vilegiatura*? It is not a thing; instead, it is a transient condition, which occurs when individuals or social groups move, mobilize, and reserve a space-time cut where the needs of practitioners will be met. That is, enjoyment/leisure/fun resides in the condition of becoming temporarily settled in a place that is not your residence, your dwelling, or your home. Generally, vacanciers seek out places to stay depending on social (festivals, cultural practices, customs) and/or environmental (beach, mountain, fresh air) conditions. They are complementary pairs, travel and the temporary stay, tourism and *vilegiatura*.

The desire to stay temporarily in villas is the product of social relationships in the city; it is a dialectical result of the potency and limitations of the way of life in urban agglomerations. Country houses, which were residences on the city outskirts, initially served the romantic desire to renew the vital forces worn down by dynamic city life. From the nineteenth century, recreation merged with modern, worldly leisure. Travel, tourism, leisure, places to visit, and spaces to inhabit temporarily became extensions of the city. If the world of work had changed, time and leisure spaces also changed. *Vilegiatura* was transformed, no longer restricted to aristocrats. Tourism and second homes became the expansion vectors of the urban fabric. Travelers took urbanization with them.

The urban society induced by the industrialization and economism of social production imposed the spatial-temporal separation of work and leisure as rational and

© The Author(s), under exclusive license to Springer Nature Switzerland AG 2020
A. Queiroz Pereira, *Coastal Resorts and Urbanization in Northeast Brazil*,
SpringerBriefs in Latin American Studies,
https://doi.org/10.1007/978-3-030-46593-3_1

logical. The changes in the social-spatial pattern of the city since the twentieth century have contributed to the redefinition of other urban processes. The desire for *vilegiatura* is one example. The formation of neighborhoods or intra-urban sectors marked by social homogenization (understood as family groups with similar income and occupation patterns) contributed to the dissemination of *vilegiatura*. This process of social cohesion is not usually due to a juxtaposed localization (geometric space), but instead by the dissemination of fashion, by a cultural value ascribed to practices and the places reserved to them.

Forged by the new urbanization, *vilegiatura* passed through the massification of desire and an established status quo. This urbanism is the fruit of Modernity. Urry (1996) refers to the experience of "being a tourist" as one of the characteristics of the modern human being. Here, the meaning of being in *vilegiatura* crosses eras and reaches Modernity, passing through a series of discontinuities. It is believed that these discontinuities are understood by examining the transformations occurring in and by the city.

Munford (2008) identifies the genetic incorporation of new practices into the city context, pointing out that "each new component of the city, for this very reason, initially appeared outside its limits, before the city" (p. 32). The dissemination of *vilegiatura* on the most varied natural sites gives continuity to the process of expansion and production of new models of territorial organization, now called the city. This pattern of incorporation crosses the centuries and different political, commercial, and urban-industrial city models. Even if *vilegiatura*'s relevance in the production process of the (non-)city is not taken into consideration, Munford makes a pertinent point.

From the time of the Roman urban model, the *villas de otium*, a generic form of *vilegiatura*, originated within the city walls. With the Renaissance (of the city itself) and the consolidation of Modernity, this urban dimension has persisted, relating to and fulfilling a historical role, the city and what it is capable of bringing together. Volochko (2008) reminds us of the possibilities of spatial-temporal practices conditioned by the production of cities taking place, citing the strong examples of philosophy and politics in the Greek cité. In this sense, the rebirth of cities promoted *vilegiatura* as an antithesis to agglomeration, linked to a countryside-city transition or interface.

The desire for the non-city and nature, for the "Champs Elysées," is added to the city's intrinsic characteristics of agglomeration, sedentarism, market, and public administration, as stated by Lencioni (2008). In itself, this desire is a mutation. In his considerations on style and social satisfaction, Lefebvre (1991) compares the aristocracy and the bourgeoisie, evaluating that while the former developed and ensured a concept of pleasure, the latter did not even taste satisfaction, let alone happiness.

The transformations in the model of society and the characteristics of urbanization, and so of *vilegiatura* and leisure, refer to the reproduction of social relationships. This process is not limited to the *stricto senso* environment of production. It is linked to society and its relationships of production, the urban, private life, and leisure.

Another common misconception is the description of urbanization as a phenomenon that develops in successive stages, driven by industry (Lefebvre 2004). In the case of coastal urbanization during the second half of the twentieth century, in several locations, fixtures and flows were installed without, however, passing through the accumulation model directly provided by industrialization. Both *vilegiatura* and later, tourism, produced (and produce) more accessible spaces and, thus can assimilate the demands derived from the near and/or distant urban and metropolitan agglomerations.

The configuration of contemporary urban spatial forms refers to a complex mix of morphologies and functions: the fragmented competes with the compact, and functional diversity prevails to the detriment of monofunctional agglomerations (Lefebvre 1999).

The global advance of the city and the urban in the twentieth century shows that *vilegiatura* follows the route of global transformations. Thus, the romantic model is reinvented, and the spaces for leisure and sojourning do not detach themselves from the urban fabric; on the contrary, they are inseparable from it. The social groups in the twenty-first century who are able (and desire) to practice *vilegiatura* perceive the natural with an urban gaze. Vilegiaturists on the coast, for example, demand and "need" a controlled, "reshaped" nature, composed of the same technical systems found in metropolises and medium-sized cities. There is a theoretical understanding of this reversal since "for many centuries the City was perceived, conceived, and appreciated regarding the countryside, but through the countryside and Nature. Now, a century later, the situation is reversed: the countryside is perceived and conceived in reference to the City" (Lefebvre 1991, p. 126). So *vilegiatura*, previously considered a rural practice, now has the potential to influence metropolization. That is, originating in an agglomeration, it facilitates the continuous or discontinuous extension of the urban fabric, contributing to the implosion-explosion process of the city. In this way, spatial fragmentation is present as a geographical facet of the process of metropolization, which includes the northeastern metropolises.

In the context of the variety of today's urban agglomerations, *vilegiatura* appropriates voids and small villages/towns, on the urban-metropolitan fringes, and even from the city's internal texture (the metropolis). In neo-capitalist conditions, the metropolis plays a significant role, it "takes on the function of command and irradiation of transforming processes, as well as being the privileged place from which the urban world is read" (Carlos 2004, p. 67). The growth of these urban forms in the last four decades, especially in "Third World" countries, has led to a reorganization of the agglomeration model, marked by new relationships between center and periphery.

In the last century, romantic urban morphology, as attested by Munford (2008), has been replaced by the modern pattern, organized according to the prescriptions of fragmentary and functionalist rationality. In this way, metropolization has become a global phenomenon linked to the new needs demanded mainly by the actions of the State, the middle classes, and business groups. It is clear that the imposition of networks on the territory has allowed new locational patterns, filling peri-urban spaces with socially valued activities, including those associated with rare natural amenities (Lefebvre 2004). These peri-urban conditions are described by Egler (2001) when mentioning the expansion of cities and the formation of large agglomerations, due to the elevation of living standards, above all, the increased consumption standards of the middle class. There has been a reallocation of several productive activities and new ways of residing, resulting in an increase in pendular mobility, through different mobility and communication technologies.

The massification of leisure practices encouraged the expansion of cities and the transformation of rural spaces into urban ones. Worldwide, beach areas are, without a doubt, the natural sites prioritized to receive this type of flow. Even seasonal visitors lack the urban infrastructure and real estate to fulfill their desires. The incorporation of the beach into the city is a nineteenth-century legend, but it gained spatial expressiveness and speed throughout the twentieth century.

1.2   The Beach and Leisure Stays, from the Nineteenth and Twentieth Centuries

In the modern western world, the beach has been symbolically transformed. Over approximately three centuries, Nature was appropriated by the artificialization of the landscape. This apparently incoherent finding demonstrates, in fact, the dialectic between the social practices of involvement/rediscovery of nature and the symbolic and material redefinition of the environment. The production of a new Nature is an indispensable condition of this process. In his classic work, Corbin (1989) defines the paths of this social process, indicating how literature, painting, medical discourse, and even new interpretations of biblical texts reformed the images attributed to the coast. At the turn of the twenty-first century, beaches and their nature are mostly "humanized," portrayed by images and practices linked to pleasure and the rediscovery of a particular model of well-being. When discussing the production of modern maritime practices in the tropics (coastal tourism and maritime tourism), Dantas (2009) identifies it as an effect of fashion. The beach, in Modernity, has become the place of fantasy, of mingling and parties, as Lambert et al. (2006) recall.

The rapprochement of European society with the effects of maritime diversion led to many new social practices. Preceding everything that we now know as tourism, the temporary stay, *vilegiatura*, has created a symbolic aura capable of transforming the coastal landscape into a "natural sanctuary" of leisure and pro-elitist social differentiation. When and under what conditions was such a phenomenon established? According to Corbin (1989), in the summer of 1735, on the beach of the English coastal town of Brighton the Rev. William Clarke "set up rural sports (…), initially shaping what we now call maritime *vilegiatura*" (p. 90). The same author underlines his classical background when he recalls that the English invention was strongly related to Roman traditions. The British reversed a logic that attributed labor functions to the countryside and beaches, that is, they were spaces of production. Amid an urban revolution, the most powerful nation in the world, at the time, adopted the "Bath" and "Spa," which initially followed a medical-therapeutic discourse, in Brighton, the first modern beach.

Although it had been practiced since the beginning of the eighteenth century, it was only at the end of the first quarter of the following century that *vilegiatura* houses "turned" to the sea. Until that point, the dominant situation had excluded the sea as a preferential space in the development of social practices during the period of *vilegiatura*.

At the end of the eighteenth century, the first spa resorts appeared in continental Europe. These were different from the British ones. In the Baltic and the North Sea, *vilegiatura* came even closer to the sea, leading to the emergence of Scheveningen (Netherlands) and Ostend (Belgium), and in France, Boulogne, Dieppe, and Biarritz. In the latter, the first villas date back to 1841. The Spanish court established San Sebastian and the Portuguese nobility Estoril and Cascais. Mantobani (1997) explains the Argentine case and characterizes the *veraneio* (summer) of the first resorts. The locations of maritime *vilegiatura* spread non-linearly throughout the continent, becoming more and more diversified, situated in places with distinct characteristics. A similar movement spread to the former European colonies. In both Argentina and Uruguay, the last quarter of the nineteenth century was marked by the formation of the first resorts.

Describing the production of coastal-maritime urban forms during the "premier âge Touristique"—1850–1930—Debié (1993) shows that the transformations in the landscape of the Mediterranean centers (Côte d'Azur, Cannes, Monaco, Antibes, Nice) began with the desire for a stay during the winter season. This resulted in the elaboration of a planned space, where the boardwalks (promenades) were a successful piece of urban planning so that the centers were attractive in the winter, rather than as seaside resorts. With electrification, and especially the rise of the automobile, as well as the inauguration of the summer centers on the coast, the promenades lost space to the coastal boulevards.

Still in the nineteenth century, new social demands produced the first bungalows in Britain (Urry 1996), contextualized by the following conditions: I. the increasing attraction of visiting the coast for more than strictly medical reasons, such as the beautiful landscapes and invigorating air; II. the increasing demand by sectors of the middle class for accommodation away from other people; III. the possibility of contemplating the sea in relative solitude; IV. the growing popularity of swimming; and V. the perception of the need for semi-private access for the whole family, especially children. There is not a single exceptional feature, both sandy and rocky beaches received vilegiaturists; it is worth remembering that the beach face only became a space for games and exposure to the sun later on. However, a fundamental social change occurred at the end of the nineteenth century, when the urban bourgeois class played a leading role in patronizing seaside resorts. Although they are not a bourgeois invention, it is the ideology of the time that founded the notion of leisure by the sea, including, under different conditions, the most varied social strata.

At the epicenter of the phenomenon, in 1914, the French maritime tourism resorts surpassed hot springs and mountain resorts in both number and attendance (Boyer 2008). *Vilegiatura* grew in the twentieth century, although there was a hiatus during the First World War. After 1919, growth restarted until 1929, when the process declined. The taste for maritime *vilegiatura* "descended" from the islands of Great

Britain to the southern tip of Europe. The theories of heliotherapy influenced this spatial change in maritime *vilegiatura*, also associated with the practice of a "change of air" (pure air) against pulmonary and respiratory diseases. Thus, beaches in lower latitudes, with a higher incidence of luminosity and warm temperatures throughout the year, gained relevance in the international division of maritime *vilegiatura* locations.

The maritime *vilegiatura* practiced in Europe during the nineteenth century was reformed in the twentieth century. In the 1920s, the taste for the sea and the sun came together, and in the second post-war period, this desire grew exponentially. This new pattern was not only based on climatic-therapeutic norms and prescriptions but also valued exposed athletic bodies. Thus, "the aristocratic notion of marble [skin] color was replaced by the bourgeois notion of a bronzed color, and on undressing the human body gained new value and a new coloration" (Correa 2010, p. 177). Debié (1993) ascribes the contemporary urban morphology to the social inclination for the summer-sea binomial. Boyer (2008) mentions that before this cultural change, many hotels and private residences in the Cote d'Azur remained closed throughout the summer.

By 1924, due to American gatekeepers, Juan-les-Pins, in France, had become a thriving resort. In the first quarter of the twentieth century, there was a noticeable change: kings (and aristocrats) gave way to other agents in the process of promoting places. One example was Hollywood artists. Brigitte Bardot popularized Sant Tropez and, in Brazil (in the 1960s), she became an icon on the beaches of Armação de Búzios, in Rio de Janeiro. In addition to American stars, in Germany, the persecution of the practices of nudism and naturalism during the Nazi and post-war period and the country's division was significant in the spread of *vilegiatura* to the Mediterranean. Southern Europe was a suitable place for German practitioners to re-establish their sunbathing and naturist colonies.

North American businesspeople and real estate developers became interested in the Mediterranean and established economic relationships and advertising to promote sunshine *vilegiatura*. Thus, Florida and California (with its film industry) were transfigured into spaces of maritime *vilegiatura*.

Related to the global economic changes of the "glorious thirties," maritime and sunshine *vilegiatura* became hedonistic and market-driven. At that time, the term resort gained the connotation of a self-sufficient hotel and leisure complex. The village became synonymous with a leisure space and a club, a society of equals. The same consumer-driven society, which invented vacations and regulated working hours, organized tourism and maritime *vilegiatura* as products that "imitate and reproduce the work of previous (pre-capitalist) societies launching them into mass consumption" (Lefebvre 1991, p. 35).

From the last quarter of the twentieth century, the beach areas of tropical countries have undergone major transformations. The phenomenon became globalized. Reports from Homes Overseas, Place in the Sun and Overseas Property, for example, analyzed and disseminated the new maritime *vilegiatura* locations. Destinations were recommended in Morocco, Turkey, South Africa, Cyprus, Brazil, Malta, and

Dubai. Spas designed for sea and sunbathing spread throughout Central America and the Caribbean, the Mediterranean, the Middle East, and Southeast Asia.

The large urban-metropolitan basins simultaneously acquired two conditions: i. spaces emitting millions of travelers; and, in the case of coastal (and tropical) locations, ii. they housed spatialities occupied by hotels, second homes, theme parks, maritime resorts, and tourists. Leisure activities came to represent the essential characteristics of these urban forms.

1.3   Coastal Metropolises and Real Estate for Vilegiatura: Massification in the Twentieth and Twenty-First Centuries

According to Hall (2014), the topics of second homes and tourist mobility have been discussed comprehensively in specialized publications in the United States, England, and Scandinavia (Ellingsen and Hidle 2013). These works include detailed case studies and regional contexts and are methodologically organized into qualitative and quantitative researches. Nevertheless, even in these countries, there are difficulties in obtaining and processing up-to-date and reliable data (Czarnecki and Frenkel 2014).

International studies correlate the ownership and use of second homes with aging populations (especially in Europe), income generation (rent), economic development (Hoogendoorn and Visser 2011), the transformation of rural or peri-urban landscapes, and the management of the seasonal flows of property owners.

Understood as a return to rural origins, the use of second homes is justified by the quest for peace, tranquility, and the construction of a "lifestyle" (Anabestani 2014; Müller 2011). Authors such as Adamiak et al. (2017) believe that the case of Finland evidences the development of a process called "counterurbanization."

Hall (2014, 2015) also considers the relevance of studies that analyze the differences between the behavior of "second residents" (referred to vilegiaturists in this book) compared with other tourists. The research corroborates the theoretical premise defended here: as urban life changes, the model of how second homes are used and produced will also change. A similar view is advocated, which establishes a simultaneous inductive relationship between second homes and the urbanization process. Regarding the differences between users of second homes and ordinary tourists, it is believed that the existing socio-cultural and geographic studies on vacanciers and vilegiatura elucidate and illuminate these issues effectively, especially in countries of the Global South or peripheral capitalism.

Following the growth of the fabric of cities in both the advanced capitalist world and peripheral (or emerging) countries from the middle of the previous century, metropolitan spaces were conceived and produced as a function of *vilegiatura* and not just dwelling places. It is important to consider that *vilegiatura* first "colonized" these spaces. Despite the association with nature, in practice, *vilegiatura* tends to reproduce urban space.

Lundgren (1974), for Canada; Hiernaux-Nicolas (2005), for Mexico; Colás (2003), for Spain; Marques (2003), for Portugal; and Dantas et al. (2008), for Northeast Brazil, have demonstrated the urban logic of the occupation of spaces promoted by *vilegiatura* and by its primary real estate expression, the domicile of occasional/seasonal use.

Hall and Müller (2004) attributed the term "recreational hinterland of an urban center" to these spaces. According to Limonad (2007), in terms of urban morphology, both the tentacular growth and the discontinuous expansion of the urban network mark the spatial archetype of the hinterlands of contemporary cities, characteristics also present in the extensions promoted by the needs of leisure spaces. In an earlier study (Pereira and Dantas 2008), this characteristic was observed in the case of the metropolization of *vilegiatura* in the Metropolitan Region of Fortaleza, Ceará, Brazil.

Roca et al. (2009) analyzed the demand for leisure in the second homes in Portugal. The authors investigated the Portuguese *concelhos* (similar to the Brazilian municipality) and established seven clusters. Among these, the three most dynamic are those related to the production of urban space (the *concelhos* of the Metropolitan Area of Lisbon, the peri-urban areas, and those of the holiday resorts). There is a diversity of typologies in these clusters, including compact and vertical spaces and horizontal and highly space consuming occupations.

In Spain, a vast bibliography focuses on the process above. Ribamontan al Mar, in Cantabria, is a well-studied example. In the early 1960s, the city became a peri-urban space devoted to leisure, mainly for the inhabitants of Santander, and there was, at the time, a predominance of speculative or investment activities (Latorre 1989).

There are other important examples of rapid transformation in Spain. Aledo et al. (2012) deal with second homes on the dynamic Costa Blanca, while Garcia-Ayllón (2015) details the production of the urban space in La Manga. In Oceania, the world-famous "Gold Coast" followed the North American model of tourist-real estate growth and is currently one of the most significant urbanized regions on the Australian coast (Dedekorkut-Howes and Bosman 2015).

Large tourism and real estate developments contribute to the emergence of new models of second homes, shared use, and changes in the users' profile. In the last decades, they have expanded to the borders of urbanized coastal zones. The territories of Turkey, Costa Rica, and the Gulf of Arabia have undergone an important process of urbanization (Akyol and Cigdem 2016; Barrantes-Reynolds 2011; Burt 2014). Coastal areas are equally attractive in Nordic countries, which have a long tradition of second homes and circular mobility (Persson 2015). On the other hand, the theoretical viability of the concept of residential tourism has been discussed in Brazil, Spain, Portugal, and Mexico, with a strong influence of authors writing in Spanish (Hiernaux-Nicolas 2018).

In all the studies mentioned above, the role of the metropolis as an inducer of transformations and processes is highlighted. The metropolis opens itself to global innovations (technical, financial, and symbolic) and, concomitantly, on a local scale, leads the reordering of peri-urban space. Davidovich (2001) recognizes the relevance of the metropolitan phenomenon emphasizing the technification of space due to the possibility of integration with the global.

Thus, it is evident that metropolization is not a single process and may vary according to geographic scales. Around the world, metropolitan spaces, as diverse as they are, command a hierarchy of places, mainly on a regional and national level. The metropolitan dynamic controls a peri-urban space modified according to twofold demands: the internal ones, generated by the social subjects in the metropolis; and the external ones, those initially captured and directed by the polarizing power of urban agglomeration. Thus, the urban morphology spreads out discontinuously, as the installation of communication/energy networks and circulation routes is a prior requirement, densifying according to the strategy of attraction or even the spontaneous increase of flows, mainly international ones.

In accordance with Patrick Mullins, Mascarenhas (2004), when studying the case of the municipalities of Rio de Janeiro, lists the intrinsic characteristics of a specific process to leisure activities that they call touristic urbanization. These are a predominance of consumption activities; strong demographic growth; precarious conditions of employment and income; and political dynamism on the part of the new social subjects (newly installed businesspersons and new inhabitants with a higher purchasing power, etc.), in the Brazilian case, the active participation of the State, and the creation of images about the places.

However, it is prudent to consider the particular characteristics of *vilegiatura* in the process of urbanization. Specifically, as seen above, real estate production and use of the second residences (sedéntarité) are noteworthy aspects.

In peripheral countries where the discrepancies are even more significant, the urban network depends considerably on the metropolises and the immediate spaces subordinated to them. Thus, as a result of the increasing demands of social agents, the metropolization process unfolds in the consumption and the reorganization of peri-urban space generated by production, circulation, housing, and leisure activities. With the consolidation of the urban expansion model, mainly in the middle of the twentieth century, *vilegiatura* outside the city limits was integrated into metropolization. When dealing with urban expansion, or the urbanizing role of *vilegiatura*, it is prudent to consider a general distinction elaborated by Santos (1994). The author mentions two dimensions of this process: the urbanization of society and the urbanization of the territory. Initially, through local demand, *vilegiatura* has a vital role in the diffusion of the modern nexus of the urban way of life to the populations where it is consolidated (urbanization of society). Later, in connection with other leisure practices (tourism, for example), it is an argument for the production of a space governed by the implementation of engineering systems (the urbanization of the territory).

When called to analyze more specific cases, the relationship between *vilegiatura* and metropolization becomes narrower, as observed in contemporary coastal metropolises. The triad—*vilegiatura*, the maritime dimension, and the city—in the case of the tropics (including Brazil), emerges as an interesting bias of analysis of the contemporary urbanization process (Dantas et al. 2008). Between metropolization and peri-urbanization, maritime *vilegiatura* is included among the leisure activities responsible for the reproduction of urban space, which has morphological, economic-land ownership, and social representations: the low density of land occupation (a predominance of horizontal occupations), the relative elevation of land prices and the value of land, and a tendency toward the social homogenization of users.

This book follows a spatial-temporal approach and examines modern maritime practices, especially vilegiatura. It offers historical and geographical explanations of both new and old uses/forms of buildings for temporary stay, leisure, and rest. This helps us to understand the current relationships between second homes, large real estate developments, and mass tourism. The spread of leisure in the coastal areas of Brazil has undoubtedly produced a complex process of urbanization of the territory.

References

Adamiak C, Pitkäinen K, Lehtonen O (2017) Seasonal residence and counterurbanization: the role of second homes in population redistribution in Finland. GeoJournal 82:1035–1050

Akyol D, Cigdem A (2016) Effects on the coastal areas of neolieral urbanization in Turkey. Int J Agric Environ Res 02(06)

Aledo A, Steen A, Jacobsen JK, Selstad L (2012) Building tourism in Costa Blanca: second homes, second chances? In: Nogués-Pedregal A-M (ed) Culture and society in tourism contexts. Bingley, Emerald, pp 111–139

Anabestani A (2014) Effects of second home tourism on rural settlements development in Iran (case study: Shirin-Dareh Region). Int J Cult Tour Hosp Res 8(1):58–73

Barrantes-Reynolds MP (2011) The expansion of "real estate tourism" in coastal areas: Its behaviour and implications. Recreat Soc Afr Asia Lat Am 2(1):51–70

Boyer M (2008) Les villegiatures du XVIe au XXIe siècle: panorame du tourisme sédentaire. Éditions ems, Paris

Burt JA (2014) The environmental costs of coastal urbanization in the Arabian Gulf. City 18(6):760–770. https://doi.org/10.1080/13604813.2014.962889

Carlos AFA (2004) O Espaço Urbano. Novos escritos sobre a cidade. Contexto, São Paulo

Colás JL (2003) La residencia secundaria en España: estúdio territorial de su uso y tendência. Tesis doctoral. Departament de Geografia. Faculdat de Filosofia i Lletres. Universitat Autônoma de Barcelona. Barcelona, Cataluña, ES

Corbin A (1989) O território do vazio. A praia e o imaginário ocidental. Companhia das Letras, São Paulo

Correa SMS (2010) Germanidade e banhos medicinais nos primórdios dos balneários no Rio Grande do Sul. História, ciência, saúde – Manguinhos 17(1):165–184

Czarnecki A, Frenkel I (2014) Counting the 'invisible': second homes in Polish statistical data collections. J Policy Res Tour Leis Events 7:15–31. https://doi.org/10.1080/19407963.2014.935784

Dantas EWC (2009) Maritimidade nos trópicos: por uma geografia do Litoral. Edições UFC, Fortaleza

Dantas EWC et al (2008) Urbanização litorânea das metrópoles nordestinas brasileiras: vilegiatura marítima na Bahia, Pernambuco, Rio Grande do Norte e Ceará. Cidades (Presidente Prudente) 5:14–34

Davidovich F (2001) Metrópole e território: metropolização do espaço no Rio de Janeiro. Cadernos Metrópole 6:67–77, 2° sem

Debié F (1993) Une forme urbaine du premier age touristique: les promenades littorales. Mappemonde 1:32–37.

Dedekorkut-Howes A, Bosman C (2015) The gold coast: Australia's playground? Cities 42:70–84

Egler CAG (2001) Subsídios à caracterização e tendências da rede urbana do Brasil. Configuração e dinâmica da rede urbana. Petrópolis: IPEA/IBGE/UNICAMP

Ellingsen WG, Hidle K (2013) Performing home in mobility: second homes in Norway. Tour Geogr 15(2):250–267

García-Ayllón S (2015) La Manga case study: consequences from short-term urban planning mass destiny of the Spanish Mediterranean coast. Cities 43:141–151

Hall CM (2014) Second home tourism: an international review. Tour Rev Int 18(3). https://doi.org/10.3727/154427214x14101901317039

Hall CM (2015) Second homes planning, policy and governance. J Policy Res Tour Leis Events 7(1):1–14. https://doi.org/10.1080/19407963.2014.964251

Hall CM, Müller DK (2004) Introduction: second homes, curse or blessing? Revisited. In: Tourism, mobility and second homes: between elite landscape and common ground. Channed View Publications, Clevedon (UK), pp 3–14

Hiernaux-Nicolas D (2005) La promoción inmobiliaria y el turismo El caso mexicano. Scripta Nova. Universidad de Barcelona, vol. IX, n° 194 (05)

Hiernaux-Nicolas D (2018) Turismo residencial: retos identitarios e imaginarios espaciais. In: Mazon T (ed) Turismo Residencial: nuevos estilos de vida: de turistas a residents. Publicacions de la Universitat D'Alacant, pp 17–30

Hoogendoorn G, Visser G (2011) Economic development through second home development: evidence from South Africa. Tijdschrift voor Economische en Sociale Geografie 102(3):275–289

Lambert D et al (2006) Currents, visions and voyages: historical geographies of the sea. J Hist Geogr 32:479–493

Latorre EM (1989) Genesis y formación de un espacio de ocio periurbano: Ribamontan al Mar (Cantabria). ERIA, pp 5–17

Lefebvre H (1991) A vida cotidiana no mundo moderno. Tradução de Alcides João de Barros. Ática, São Paulo

Lefebvre H (1999) A revolução urbana. Tradução de Sérgio Martins. EdUFMG, Belo Horizonte

Lefebvre H (2004) O direito à cidade. Tradução de Rubens Eduardo Frias, 3ª edn. Centauro, São Paulo

Lencioni S (2008) Observações sobre o conceito de cidade e urbano. Geousp - Espaço e Tempo, São Paulo 24:109–123

Limonad E (2007) Urbanização dispersa mais uma forma de expressão urbana? Formação, Presidente Prudente-SP 1(14):31–45

Lundgren JOJ (1974) On access to recreational lands in dynamic metropolitan hinterlands. Tour Rev 29:124–131

Mantobani JM (1997) Notas sobre el problema de la creación de los primeros balnearios argentinos a fines del siglo XIX. Scripta Nova. Revista Electrónica de Geografía y Ciencias Sociales 11

Marques TS (2003) Dinâmicas territoriais e as relações urbano-rurais. Revista da Faculdade de Letras – Geografia. I série, vol XIX, pp 507–521

Mascarenhas G (2004) Cenários contemporâneos da urbanização turística. Caderno Virtual de Turismo 4(4):1–11

Müller DK (2011) Second homes in rural areas: reflections on a troubled history. Nor Geogr Tidsskr 65(3):137–143

Munford L (2008) A cidade na História: suas origens, transformações e perspectivas. Tradução de Neil R. da Silva, 5ª edn. Martins Fontes, São Paulo

Pereira AQ, Dantas EWC (2008) Veraneio marítimo na metrópole: o caso de Aquiraz-CE. Sociedade & Natureza 20(2):93–106

Persson I (2015) Second homes, legal framework and planning practice according to environmental sustainability in coastal areas: the Swedish setting. J Policy Res Tour Leis Events 7(1):48–61. https://doi.org/10.1080/19407963.2014.933228

Roca MN et al (2009) Expansão das segundas residências em Portugal. In: Anais do 1° Congresso de Desenvolvimento Regional de Cabo Verde. Cabo Verde, pp 2448–2474

Santos M (1994) Tendências da urbanização brasileira no fim do século XX. In: Carlos AFA (orgs.). Os caminhos da reflexão sobre a cidade e o urbano. Edusp, São Paulo, pp 17–26

Urry J (1996) O olhar do turista: lazer e viagens nas sociedades contemporâneas. Tradução Carlos Eugênio Marcondes de Moura. Studio Nobel/SESC, São Paulo
Volochko D (2008) Sociedade urbana e urbanização da sociedade: elementos para a discussão sobre a problemática da cidade contemporânea. Cidades 5(8):215–242

Chapter 2
The Metropolises Grow Towards the Sea, Northeast of Brazil

Abstract The regional planning model for the Northeast and the drive for modernization and economic growth adopted in Brazil selected state capitals as hubs, leading to the diversification of the phenomena related to coastal urbanization (related to residing, *vilegiatura*, and tourism). The global process of adding value and the valuation of urban and metropolitan seafronts of the tropical coasts intensified this conjuncture. This chapter describes the analysis of the genetic components of the coastal urbanization process in the Northeast of Brazil, describing the socio-spatial variants responsible for the organization of the seafronts of the cities of Salvador, Recife, Fortaleza, and Natal, as well as the metropolitan coastlines associated with them.

Keywords City · Beach · Urbanization

2.1 Cities Become Coastal: Urbanization Moves Toward the Beach

The last two decades of the nineteenth century were the start of continuous population growth in the Brazilian coastal state capitals. However, at the beginning of the last century, the relationships with distant cities were tenuous, and overseas relationships (with Europe, above all) played a more significant role in the changes in the large cities of colonial origin. At that time, the ports were the central point where the evolving cities intersected with the beach. As stated by Dantas (2006), Brazilian tropical cities were coastal, but not maritime. In his evaluation of tropical locations, the author demonstrates that until the nineteenth century, the relationship between society and the sea (and the beach) was predominantly mediated by traditional maritime practices, namely, those that emphasized work, the transportation of cargo/people and territorial defense, but did not involve recreation or leisure.

The production of the urban and northeastern urban society advanced with the accumulation of wealth in the capital cities. Exchanges with Europe were not only material, but a civilizational pattern also emerged according to the dialectic of social actions. As total assimilation of a given model was an impossibility, practices close to

© The Author(s), under exclusive license to Springer Nature Switzerland AG 2020   13
A. Queiroz Pereira, *Coastal Resorts and Urbanization in Northeast Brazil*,
SpringerBriefs in Latin American Studies,
https://doi.org/10.1007/978-3-030-46593-3_2

Western ones were adopted (Op. Cit.). The elite, accustomed to large farms, slaves, mills, livestock, and cotton, perceived the city as a new place, based on elegant, glamorous fashions, and practices explained by rationalist discourse. Therefore, the urban pattern of the cities was redefined toward a geometric-rational pattern and related to a hygienist discourse (Costa 2006). Consecutively and concomitantly, the first industrial establishments were set up, and the tertiary sector was diversified. New services and goods redefined the daily life of the northeastern city: the streetcar, street lighting, the train, and the press, among others. A growing number of farmers moved to the city and visited their farms in the harvest periods. As was already the case in Salvador/BA and Recife/PE, the other northeastern capitals were prominent in the innovation of social practices and started to lead the territorial organization of the provinces (future federative units from 1889).

At the end of the nineteenth century and the beginning of the twentieth century, the northeastern cities showed signs of modernization linked to European models of civilization and urban agglomeration. As an example, Vasconcelos (2002), Costa (2007) list the main innovations installed in Salvador and Fortaleza, respectively. In the capital of Bahia, the law faculty was established in 1891; the electric tram was first used in 1897; in the same year, the Polytechnic School was founded; in 1901, the first car paraded through the streets; in 1903 and 1905, respectively, electricity and sewage services were implanted. In the Fortaleza, the water channeling system (1863), the first train line (1873), spinning factories (1895), and the Law faculty (1903) came into existence. Also, the José de Alencar Theatre (1910), electric trams (1913), and electric energy for residential lighting (1914) were among the new services founded.

Physical treatments at the seaside, both by immersion in the waters and because of the qualities of the pure air, fit into this perspective of the modernization of the urban society of the Northeast (Costa 2006). When compared to the process of high-income housing concentration on the seafront in Santos/Guarujá and Rio de Janeiro, the Northeastern developments are more recent. While the 1920s were the starting point for the former, in the latter, the process began in the second half of the twentieth century. Villaça (2001) attributes this temporal difference to three conditions: (a) the conservatism of the northeastern aristocracy; (b) in the case of Recife and Salvador, the lack of attractive beaches near the city centers; and (c) the late development of tourism activities in the northeastern capitals, especially Recife and Fortaleza. When outlining these constraints, it seems that the author above has not considered the insertion of the Northeast and its cities into a new territorial division of labor and power. The strength of the economic cycles and overseas contacts in the South and Southeast regions at the beginning of the twentieth century were much less evident in the Northeast. Coffee cultivation, central public policies, and the first industrialization cycle occurred in the Southeast, which was the preferred subspace for innovations.

As previously stated, foreign visitors encouraged a taste for seaside recreation. In the early nineteenth century, the Dutch Quirijin Maurits Rudolph Ver Huell, accompanied by English travelers, reported the pleasure of bathing in the mornings in the sea of Itaparica, on the coast of Bahia (Correa 2010). These pioneering initiatives led to gradual changes that introduced the recreational use of the urban beaches of the northeast, a fact punctuated by Dantas (2002) when analyzing the case of Fortaleza.

The 1920s and 1930s attest to the success of maritime *vilegiatura* in the capital cities. Far removed from the socio-spatial configuration of northern European seaside resorts, or the Cote d'Azur, the seaside resort was able to associate, in space-time, other modern maritime practices, such as sea bathing and walking, which, in isolation, had not urbanized other stretches of the coast. This association justified the selection of coastal segments suitable for the development of occasional-use housing.

The literary authors Gustavo Barroso (Ceará) and Clarice Lispector (Ukrainian settled in Rio de Janeiro) demonstrate how writers interpreted and experienced the emergence of the taste for beach sojourns. The first describes Fortaleza's coast in the early twentieth century and praises the landscape: the emerald sea, the white sands, and the strong presence of the fisherman and his raft. The second author, who lived in Recife from 5 to 15 years of age (1925–1935), describes with enthusiasm, in the chronicle *Banho de Mar*, her joy on the days when she and her family woke up early, took a tram, and headed for the beach in Olinda to bath in the sea.

If sea bathing was part of the daily life of the urban populations of the Northeast from the end of the nineteenth century, the regional bibliography shows that the 1920s were the beginning of the urbanization process of the beach for leisure. From that time, alongside country houses in the city outskirts, "beach houses" represented the urban expansion of the coastal capitals. In Salvador, the elites selected the Atlantic shore for modern maritime practices, specifically in 1923 with the construction of subdivisions in the spa city of Amaralina. Rio Vermelho, and later Itapagipe were incorporated in this process (Mello e Silva et al. 2009). The construction of access roads and the availability of land for sale promoted the Atlantic shoreline, opening the way for future verticalization, according to Vasconcelos (2002) (Fig. 2.1).

Despite the symbology and success of the Capibaribe River, Recife and its elite also "surrendered" to sea bathing as a recreational practice. The southern sector of the city, polarized by Boa Viagem, became the location of maritime *vilegiatura* in the capital of Pernambuco. This change is significant as at the time Recife led the modernization process of urban spaces in the Northeast. In the nineteenth century, the western suburbs bordered the Capibaribe and were the dominant residential and leisure space of the city's elites. By the dawn of the twentieth century, the occupation of the southern sector (Boa Viagem and Pina) showed the ascendance of the maritime in the city over river bathing. In 1858, the San Francisco train gave access to Boa Viagem for those coming from the center of Recife. The train was a prerequisite for the subsequent trolley line connecting the train station to the beach itself, at the end of the nineteenth century. Recife pioneered the innovations in the Northeast related to the occupation of the coast by the modern maritime, as evidenced by the pioneering construction of a five-kilometer-long seafront promenade in the first half of the 1920s (1922–1926). Debié (1993) regards promenades, and avenues by the sea as singular forms of urbanization, urbanism, and urbanity produced initially between the years 1850 and 1930 in European cities. They relate to the articulation between dwelling and leisure spaces and have undoubtedly spread to the coastal capitals of the Northeast (Fig. 2.2).

The spread of maritime *vilegiatura* was not confined to the great capitals of the colonial period (Salvador and Recife). In Fortaleza, Iracema beach (formerly Peixe

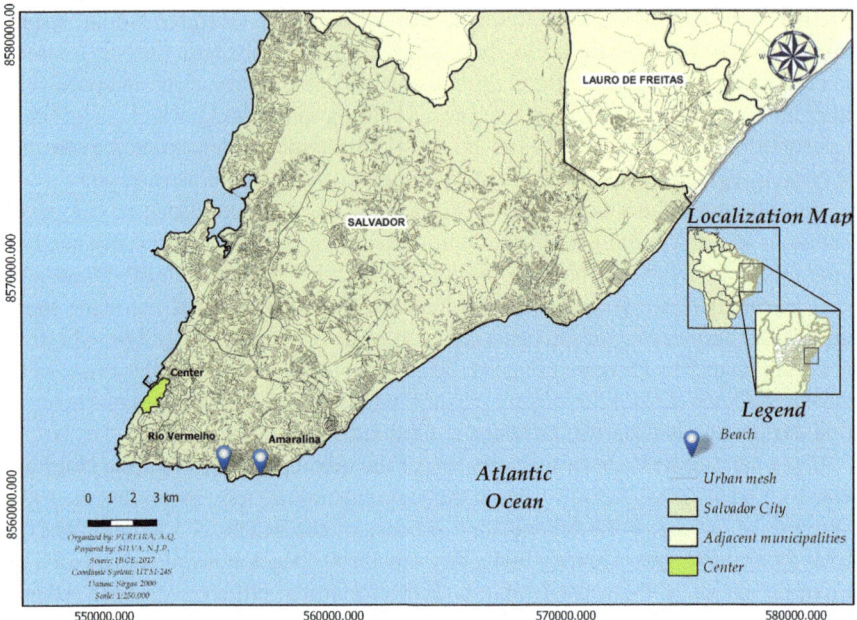

Fig. 2.1 The original location of maritime *vilegiatura* in the city of Salvador

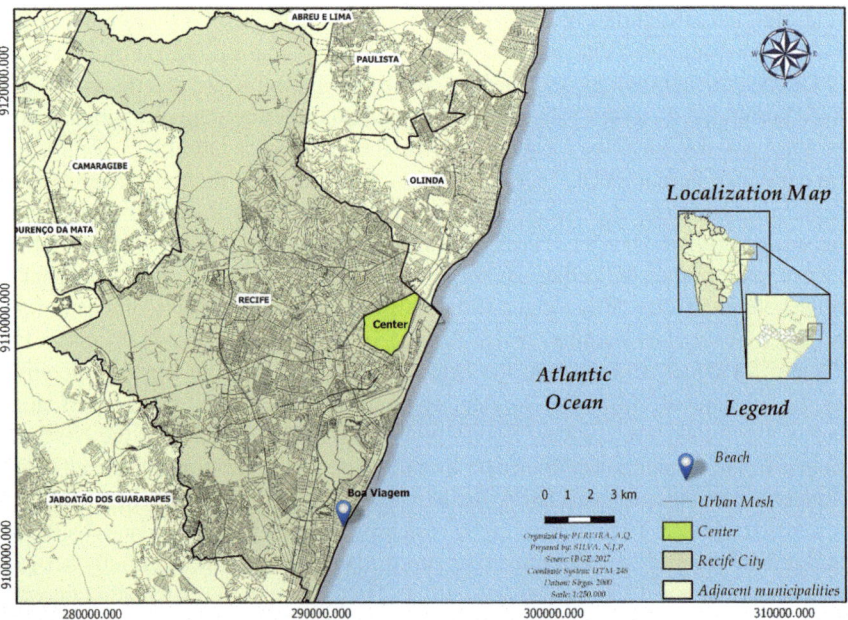

Fig. 2.2 The original location of maritime *vilegiatura* in the city of Recife

beach) was the first in Ceará used by Fortaleza's elites as a leisure space. The first temporary-occupancy residences and clubs destined to seaside leisure were built next to the well-known Metallic Bridge. The construction of the Mucuripe port further to the east caused the erosion of Praia de Iracema beach, contributing to the migration of the wealthy classes to Meireles Beach, which after the 1940s became a space of *vilegiatura* (Dantas 2002) (Fig. 2.3).

In Natal, in the decade of 1910, the Areia Preta and Meio beaches (to the south of the mouth of the river Potengi) were selected by the local elites for the development of modern maritime practices. They built occasional-use residences and in the 1920s, an improved second generation of secondary dwellings was already in place. In 1925, the Atlantic and Circular avenues were built in the form of Boulevards, linking the elite's residential neighborhoods (Petrópolis and Tirol) to the beach areas (Silva 2010) (Fig. 2.4).

In the first half of the last century, the incorporation of beaches into the capitals' urban fabric connected with a new rationale that directed the population and the spatial growth of the cities. The original occupation, the historic center, was not sufficient for the new urban model. Trams, roads, migrations and, later, automobiles, as well as the taste for the sea, contributed to the implosion-explosion of the city (Lefebvre 2004).

This process is not a copy of what happened in Europe or the capitals of Southeastern Brazil, as the influence of industrialization was much weaker. It was a movement produced primarily by a new mentality of the elite, who perceived the coastal areas

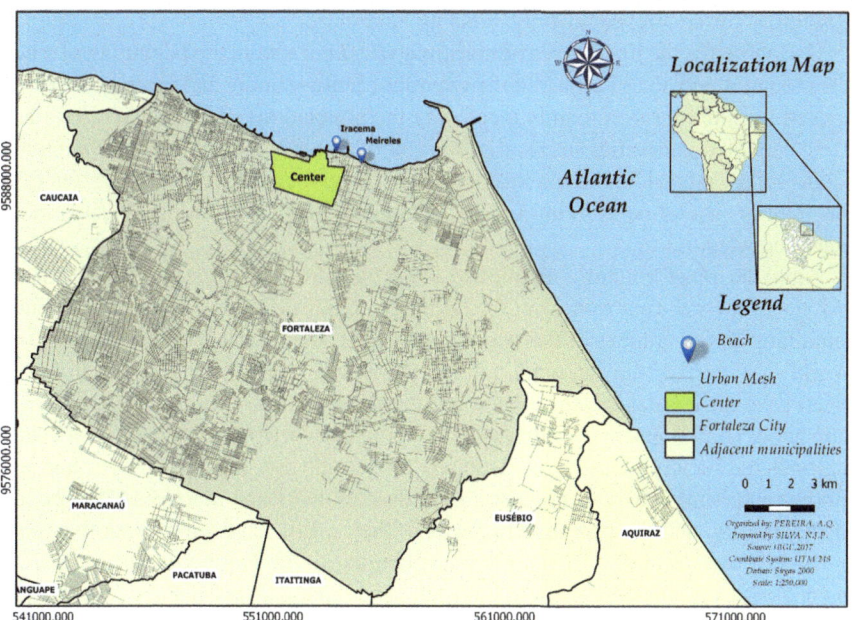

Fig. 2.3 The original location of maritime *vilegiatura* in the city of Fortaleza

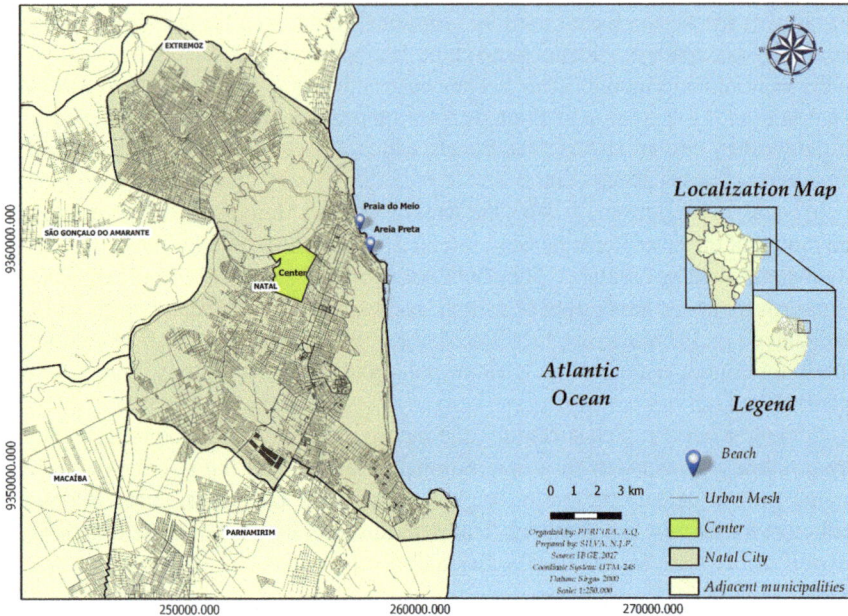

Fig. 2.4 The original location of maritime *vilegiatura* in the city of Natal

(not inhabited by the poor or occupied by other less sophisticated activities) as a possibility for leisure, exclusivity, housing, and more.

Undoubtedly, the first spaces of maritime *vilegiatura* were the beginning of a new socio-spatial division of cities. In the twentieth century, more and more poor people arrived from the *sertões*, mainly increasing the populations of Recife and Fortaleza. They occupied interstitial spaces that were not favored by the elite. The continental suburbs, the first residential spaces outside the center, began to face competition from the "balmy" shoreline as living spaces, conquered and conceived, primarily, due to maritime *vilegiatura*.

Amid the transformations mentioned above, with lesser or greater urbanistic intensity, the taste for the maritime occurred simultaneously in all the northeastern capitals, remaining in the context of the way of life of all social strata. In the second post-World War period, the internationalism of modern practices (including maritime ones) gave the coastal location a strategic character, both for its ability to receive or ship goods (traditional maritimity) and for its inclination to meet the massive needs of a globalized society focused on leisure and/or seaside housing. These were not only modern maritime practices but also urban modern maritime practices. The process of redefinition and the profound changes in the seaside have entered another stage in the last two decades. Coastal tourism planning in the Northeast has entered the agenda of development policies at both state and federal levels, and the capital cities have clothed themselves with the positive imagery of the semi-arid, assimilating their receptive tourism "vocations" (Dantas 2010).

The invention of coastal spaces as a scarce residential resource in the cities has been determined by the voracity of the potentates (who assimilated the re-signification of the sea and the maritime), the real estate market (and its projects), and mainly the municipality (formed by the former social sectors). The latter intervenes by artificializing the seaside and legislates to fragment and discriminate the functions of intra-urban spaces. In this sense, we agree with Linhares (1992) that the production of the modern beach in the tropics is the result of "a subtle game between planning and spontaneity, between actions from above and actions from below" (p. 207).

The phenomenon of *vilegiatura* (and second residences) and later the beachfront dwelling—understood as a symbol of the modernization of cities—is described, with variations in emphasis, in the specialized studies of the historical production of coastal cities in the Northeast. In relating the expansion of the capitals, these studies invariably refer to the oceanic seafronts as places of intense real estate, economic, social, and cultural dynamics.

According to Vasconcelos (2002), in 1958 it was possible to distinguish the differences in the distribution of social strata across the territory of Salvador. The wealthy classes lived near the beach, preferring to settle in places such as Graça (more continental) and others on the seafront (Vitória, Barra, Barra Avenida), continuing along the Atlantic beaches, influenced by the conclusion in 1949 of the road along this sector. After this, the real estate market continually parceled the space, transforming it into low-density residential urban land, with the edge of the bay destined to the lower classes.

In 1968, the Salvador City Hall passed the Urban Reform Law, advocating the removal of the lower class housing along the seafront, reserving these areas for tourist activities (Carvalho et al. 2004). In the last three decades of the last century, the Atlantic coast of Salvador (from Farol da Barra to Stella Maris) has been incorporated into the dynamics of leisure and housing. Hotels and skyscrapers replaced single-family homes, making it a new centrality in the city of Salvador. According to Vasconcelos (2002), the road system planned for the city contributed significantly to this phenomenon, since it connected the main existing corridors to the Atlantic corridor. In the 1980s, private initiative and the municipal and state-level public authorities invested in the construction of Paralela Avenue, the Bahia Administrative Center, the new Bus Station, and the Iguatemi Shopping Mall. The location selected for all these investments continued to induce the city's expansion toward the northern shore. Throughout the 1970s and 1980s, from Barra to Itapuã, urban plans and municipal interventions consolidated Salvador's ocean shore as a space valued by city marketing and the wealthy strata's preferred housing location, according to Sousa (2010).

In the first half of the twentieth century, in Boa Viagem (Recife), basic infrastructure systems (sanitation, street paving, water supply) were incorporated into the natural environment of a fishing village. The fragile occupancy of *vilegiatura* promoted by the townspeople accelerated due to the parceling of the land in 1940. The arrival of the US military in the Northeast during World War II was another strong influence, especially in the cities. The foreigners standardized the taste for the sea and sun and selected Boa Viagem as their living space. These transformations converged with urban legislation passed in 1953, rezoning this area of the seafront as urban. Concerned with understanding the formation of urban land value in the capital of Pernambuco, Alves (2009) outlines the time frame and describes the main changes in the seashore in question. In the 1970s, when the urban area of the region was multiplying, Boa Viagem competed with the historic center, concentrating new services, and specialized trades. The real estate market expanded to meet the sophisticated demands of the wealthy classes residing there.

In the case of Fortaleza, Souza (2006) describes the growth of the Aldeota neighborhood and its popularity with the elite, demonstrating that its proximity to the eastern seafront was a relevant factor in the increase in land prices due to expansion commercial-residential demand. This fact can be observed by the constant construction of buildings replacing the old one-story houses. The choice of the city's eastern seafront intensified in the neighborhoods of Praia de Iracema and Meireles. To these changes, Costa (2007) adds the migration of leisure, since "existing clubs in the city center were transferred to areas nearer the sea that become leisure and housing options" (Op. Cit., p. 71). The year 1962 was pivotal in this process. As part of the Fortaleza Master Plan, the urban planner Hélio Modesto designed and began the construction of the Beira-Mar Avenue, similar to the avenues along beaches in the Southeast. In the following decades, public and private actions continued to modernize the city's seafront, transforming it predominantly into a leisure space. At the end of the 1970s, the first boardwalk was built on the seafront, bordering the Beira-Mar Avenue. The Land Use and Occupancy Law of 1979 advocated verticalization along the avenue, thus replacing the houses with skyscrapers. In the 1980s, other boardwalks were built on the Iracema, Futuro, and Leste Oeste beaches. In the 1990s, Praia de Iracema went through many changes: the old Cassino Estoril and the Ponte dos Ingleses were restored. The pier received a shop for the observation of cetaceans, kiosks, a memorial with photos of the history of the structure's construction, and a visitor's area, to watch the moon and the sunset. In the year 2000, the public authorities intervened in the same area again, this time creating a landfill, which extended 1100 m and widened the beach by 100 m.

The case of Natal (RN) demonstrates the interaction between local and distant influences. As in other coastal cities, society started to modernize the city in the first decades of the twentieth century, and in 1939–1980, this culminated in the incorporation of the seafront furthest from the historical center into the urban fabric. The geostrategic function of the city resulted in a flow of temporary (mainly North American) residents, a fact that contributed to the dynamization of the local real estate market and redefined the city's provincial characteristics (Silva 2010). In the 1970s, Ponta Negra beach, the southern end of Natal's coastline, was transformed into a leisure space, mainly by maritime *vilegiatura*. However, there was an enclave between the beaches of Areia Preta/Meio and Ponta Negra: a military zone, formed by the coastline and dunes, later named Parque das Dunas (legally established in 1977). The zone was an obstacle to the formation of a seafront landscape similar to those along the beaches of Boa Viagem and Meireles Beach (an avenue by the sea and continuous verticalization parallel to the coastline). To overcome this restriction, the Via Costeira/Parque das Dunas project was built in 1983, it was mostly a rapid transit road (in the form of an avenue) approximately 12 km long, linking Ponta Negra to Pino beach. Subsequently, encouraged by public policies, the construction of hotels was concentrated in this space, adapting the city to the same model as the other northeastern coastal cities. Linked to the modern maritime social phenomenon, this planning restructured the city, which stands out in the northeastern scenario. Silva (2010) considers this intervention as the start of a new intra- and interurban context assimilated by Natal due to "sun and sea" tourism. The densification of leisure structures occurred in Ponta Negra, where residential buildings, single-family units, inns, and hotels are found side by side.

It is possible to cite examples of the other northeastern capitals on the coast. João Pessoa (PB) was organized from a center located far from the seafront. However, in the last decades of the twentieth century, there was a similar movement that described so far in the other large northeastern agglomerations. The seafront, home today to the Manaíra and Tambaú neighborhoods (Bezerra and Araújo 2007), was transformed from a sparsely occupied area at the beginning of the twentieth century into a subcenter dynamized by the actions of the real estate market and the demands of the high-income population, who want to be close to the maritime.

The maritime *vilegiatura* associated with the ascendancy of the *vilegiatura* of the sun has led to a new international division in the consumption of the littoral zones due to leisure. In this context, due to their natural locations, the northeastern capitals have emerged as spaces potentially available to globalization. The high-income classes, now influenced both European and international (mainly North American) standards, adopt a model of use of the seafront, consolidating the fashion of living by the seashore. This process only appears to clash with *vilegiatura*. The consolidation of the change of use according to residence has established a new social content, also linked to status and scarcity, which disseminates the desire for a "place on the beach" to the other social strata. This socio-cultural premise conditions real estate dynamics (market relations and urban planning legislation) and regulates the private ownership of coastal land. Thus, alongside socio-cultural valuation, this added value conditions and restricts the maritime quest to specific groups. As well as residential buildings, hotels, and other leisure establishments favor this part of the city.

The port function, low-income housing, and fishing and extractive activities are still part of the spatial mosaic of the seafronts of the largest northeastern coastal cities. However, since the 1970s, public and private projects and actions aimed at removing or redefining traditional functions have been commonplace. The collectivity has assimilated the images associated with modern maritimity. For example, projects for the "requalification" of port areas have been proposed, which include replacing shipping containers with the structures needed to attract cruise liners or even the construction of marinas. Urban seafronts have become the icons of the cities' modernization, opening themselves to the various facets of internationalism (architectural standards, companies, foreign users, leisure equipment, and circulation models).

2.2 The Dawn of the Process of Coastal Metropolization, Twentieth and Twenty-First Centuries

Amidst the intra-urban transformations on the fringes of northeastern Brazilian cities in the 1970s, the scale of the maritime urban phenomenon widened. This was the start of the formation of what we call the peri-urban space of maritime *vilegiatura* and leisure, a situation found in other metropolitan coastal centers.

At that time, urban planning policy conceived a territorial organization based on a metropolitan format. Polarized by the capitals, which were already the most dynamic cities, the adjoining and dependent municipalities were grouped into two categories: those that considered future integration actions and those that, before institutionalization, shared spatialities produced and/or controlled by capital (industrial districts, housing estates). However, studies of metropolitan spaces are directed mainly to the spatiality of production, without addressing the spatiality of reproduction. The spaces of maritime *vilegiatura* in the municipalities neighboring the capital city, produced by the demands of city dwellers, are included in this study.

When dealing with the institutionalization of metropolitan regions in Brazil, it is important to mention how they were constituted. According to Davidovich (2004), in the 1970s, the military dictatorship envisaged the institutionalization of a "hierarchical aggregate of cities, functionally interdependent, [which] represented a basic resource to meet the common goals and principles of the equilibrium of the system" (p. 198). However, the first nine metropolitan regions (MR) differ significantly from each other, since they concentrate the socio-spatial characteristics of their regions and states. For the MRs of Salvador, Recife, and Fortaleza, the public actions promoted by SUDENE made them privileged hubs in the allocation of resources and/or fiscal incentives primarily to consolidate industrial activity with the spatialization of the industrial districts. Nevertheless, the processes are not limited to the secondary sector, and neither is the metropolitan municipalities' distribution of spatialities equitable. Davidovich (2004) reflects on the central nucleus and characterizes the Brazilian scenario. Due to the vertical and concentrating nature of planning, the MR's nuclei are the most favored, reproducing relationships of control over their space of influence on a metropolitan scale.

For the northeastern metropolises, the process of metropolization is not primarily defined by the cooperation between the municipalities, but by the production of spatialities engendered by the overflows and needs generated in the central nucleus. Leisure on the coast, which includes maritime *vilegiatura*, although not included in the core of strategic actions, remained in the interstices and provided the (re)production of urban space, forming metropolitan spatiality (even before the institutionalization of metropolitan regions).

Planning for leisure activities on the northeastern metropolitan coast became effective in the 1990s when the Tourism Development Programs in the Northeast invested hundreds of millions of dollars in the production of a more fluid space capable of inserting the Northeast into the world circuit of international visitors. The developments of this new occupation are synthesized by Dantas et al. (2010). In this context, maritime *vilegiatura* was indirectly redefined by the insertion of the allochthonous model, especially in metropolitan spaces.

Globally, the population of large urban agglomerations is primarily responsible for the production of spatial leisure spaces. This is intensified when extensive coastal areas are part of the metropolitan context. Beach and metropolis interrelate, (re)producing a mechanism of urbanization that can be called the metropolization of marine leisure. Over time and space, the primary city brings together riches and innovations, and its population distributes the taste for modern and urban maritimity. One of the first results is the dissemination of basic infrastructure networks (in the Northeast, mainly terrestrial transport routes and the electricity network) fomented by planning at the metropolitan level marked by the political discourse and economic agents based in the metropolitan nucleus.

Ruled by the four core cities (Salvador, Recife, Fortaleza, and Natal), another 29 coastal municipalities make up the metropolitan regions in the Northeast, where a greater diversity of new models of maritime *vilegiatura*, tourist complexes, and other means of lodging and leisure are detected. With ten and nine, respectively, the MR of Salvador and MR of Recife have the highest number of this type of municipality. For the former, 80% of the municipalities (of the total of 13) are coastal; in the case of the second metropolitan region, the percentage is equal to 60% (out of the total of 15). The importance of the coastal environment for the metropolitan regions is evident, as in all cases the expansion process was guided by the insertion of coastal municipalities. This is because modern maritime practices (*vilegiatura* and tourism) have led to an appreciation in the value of coastal spaces, not only symbolic and imagistic but, above all, the financial value of the metropolitan coastal location. Traditional maritime practices of fishers and other coastal villagers are associated with the real estate demands of the metropolitan vilegiaturists, followed by non-native and international entrepreneurs and vilegiaturists.

In addition to the functions of ports (Ipojuca-PE and São Gonçalo do Amarante-CE), air transport (São Gonçalo do Amarante-RN), industry (Camaçari-BA and São Gonçalo do Amarante-CE), and housing, leisure activities in coastal municipalities occupy a strategic location in the formation of the respective metropolitan spaces.

Regarding the length of the coastlines, the metropolitan coast of Salvador is the longest, at 227.5 km; Fortaleza, Recife, and Natal follow with 198.6 km, 141.5 km, and 102.4 km, respectively.

These geometric characteristics, together with the road access system, also determine the level of integration of the coast to the hub city and greater or lesser availability of real estate. In order to evaluate the availability of areas according to modern maritime practices, the particularities of natural sites and the presence of socio-spatial practices, including traditional maritime ones (port, fishing villages, environmental protection areas) and industrialization, should be taken into consideration.

According to Almeida (2006), in the case of Bahia, Salvador's relationship with its environment has been redefined. The southern region, marked by traditional industry, lost space to the northern bay area (São Francisco do Conde, Candeias, Simões Filho) and to the northern coast (Lauro de Freitas and Camaçari), spaces marked by the presence of a new industrial model. However, the same author recalls the construction of the Estrada do Coco (1975) and the Linha Verde (1993) on the north coast, a road infrastructure that enabled "the multiplication of summer properties and tourist facilities, including large international resorts, along the coast" (op cit, p. 29). Silva et al. (2009), Carvalho et al. (2004) highlight the growing importance of Lauro de Freitas, together with Salvador, in the formation of a metropolitan spatiality. Located on the north coast, adjoining Salvador, the municipality absorbs the housing demands of high- and medium-income strata from both the Capital and Camaçari. In this case, highly valued beachfront housing has consolidated in Lauro de Freitas, mainly in the form of high-end condominiums. Even in Camaçari, remembered primarily for the Petrochemical Complex, the spatiality of the seafront is (re)produced according to the dictates of modern maritime practices, a situation evidenced by the fact that there are "allotments and leisure and tourism enterprises for the upper and middle classes" (Op. Cit, p. 283).

On the coast of the bay area, the construction of the Funil Bridge and the creation of the ferryboat contributed to the mainland's road connection to the Island of Itaparica, facilitating the dissemination of autochthonous *vilegiatura* on the Island (including the municipality of Vera Cruz). Madre de Deus, emancipated from Salvador in 1989, became part of the Metropolitan Region. Three sets of activities support its economic base: the first, the traditional base, is characterized by extractivism (fishing and seafood); the second is linked to activities to support the oil industry (Petrobras' maritime terminal); and the third, associated with leisure and autochthonous (metropolitan) *vilegiatura*.

The last coastal municipality incorporated into the metropolitan region of Salvador, in 2008, was Mata de São João. In this case, the activities that indicate a metropolitan process are the maritime leisure activities implemented after the 1990s, especially the tourist-hotel complexes of Costa do Sauípe and Forte beach (Fig. 2.5). Heavily visited beaches and resorts exist along the coast of Camaçari and Itaparica.

In Recife, the metropolitan municipalities of Olinda, Jaboatão dos Guararapes, and Paulista have a high level of integration with the hub city (Clementino and Souza 2009). As in other metropolises, the 1970s represented the milestone in the production of modern coastal spatiality in the region. According to Miranda (2008), during this period great transformations took place side by side: there were plots for sale near Boa Viagem on the seafront in Jaboatão dos Guararapes, and on the other hand, on the north coast, new neighborhoods were built in Olinda.

During the 1980s and 1990s, as well as being the space for maritime *vilegiatura*, the coastlines of these municipalities became the preferred residential location of the medium and high strata, dominating the urban morphology of Recife. Located in the extreme north of the MR of Recife, the Island of Itamaracá is an original part of the metropolitan region. Leisure and tourism activities govern their metropolitan dynamics. According to Assis (2001), the metropolitan plans and projects produced an insular landscape that is conducive to the activities mentioned above, transforming it into a "peripheral leisure zone." This is an example of autochthonous *vilegiatura*, on a metropolitan scale, with the widespread development of activities recognized by French experts as leisure of proximity, based on "temporary sedentariness."

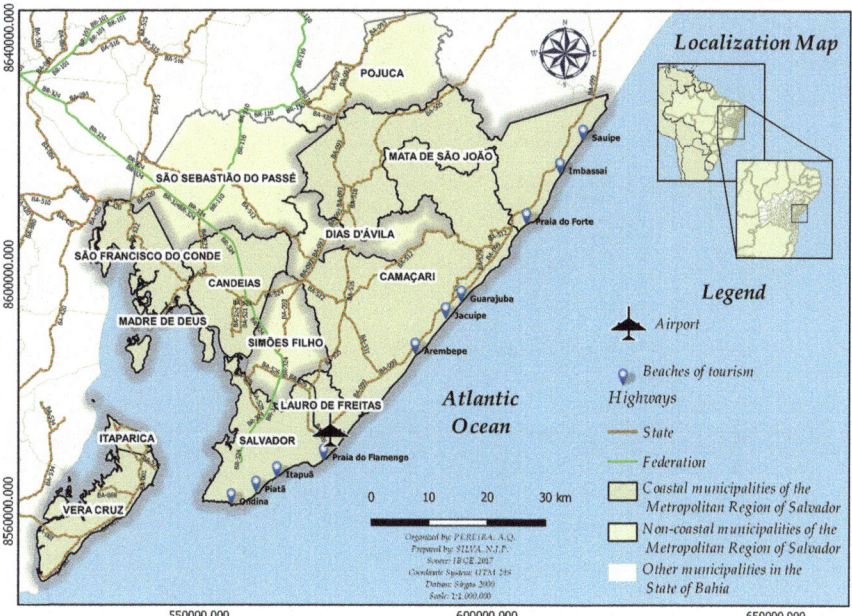

Fig. 2.5 Metropolitan coast of Salvador

The municipality of Ipojuca, the last coastline integrated into the metropolis, houses the port structures of Suape, a fact that conditions its inclusion in the metropolitan area. However, before the port, *vilegiatura* was a recurrent practice in this municipality (the 1960s), mainly in the beach of Porto de Galinhas. From the mid-1990s to the present, the number of tourist and hotel facilities has grown in those coastal places, confirming, in addition to the port, a contribution of modern maritime practices to metropolitan dynamics (Fig. 2.6). In the coastal stretches of the municipality of Cabo de Santo Agostinho, there are large-scale tourist-residential developments and complexes, for example, the Reserva do Paiva project and those located on the beach of Muro Alto.

In the case of the MR of Fortaleza, in 1972, the Integrated Development Plan for the Metropolitan Region of Fortaleza (PLANDIRF) was published. The document established zoning, indicating the city's functional relationship with the adjoining municipalities. It was concluded that the coastal areas of the municipalities of Caucaia and Aquiraz were essential and strategic in the development of leisure activities by the *fortalezenses*. After more than two decades of institutionalization, in 1999, the Municipality of São Gonçalo do Amarante (West of Fortaleza) joined the metropolitan polygon. Its insertion was justified by the construction of the Industrial Complex and Port of Pecém, a situation very similar to that in Ipojuca (Pernambuco). However, the activities of the secondary sector do not monopolize the municipal seaboard. In other coastal localities (Colônia and Taíba), autochthonous and allochthon *vilegiatura* predominate. In the case of the integration of Cascavel (East of Fortaleza),

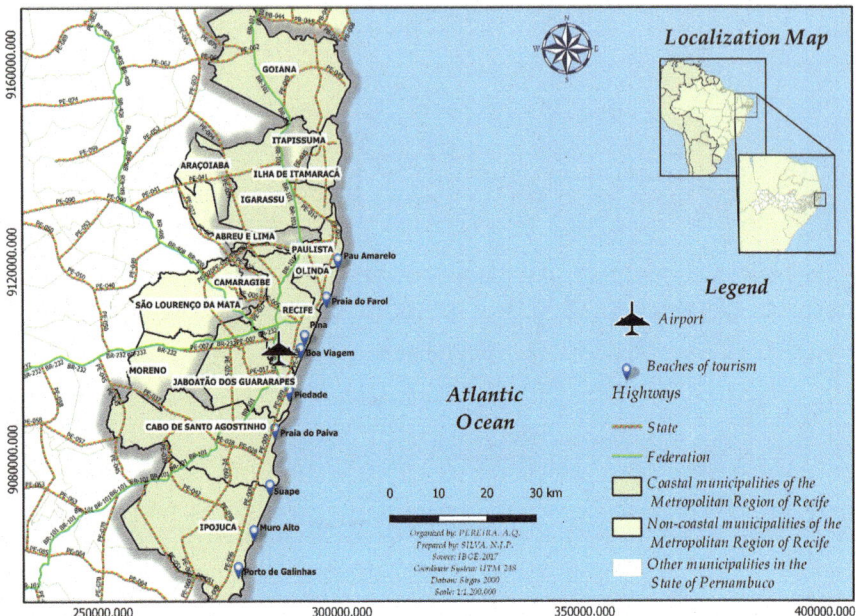

Fig. 2.6 Metropolitan coast of Recife

the only justification is the strategies for the expansion of modern maritime practices to those areas (Fig. 2.7). Aquiraz is the coastal metropolitan municipality with the most extensive leisure infrastructures, including the Aquiraz Riviera Complex (with an 18-hole golf course) and the Porto das Dunas Spa (which includes the Beach Park Water Park). Cumbuco in Caucaia is also a popular space for tourism and maritime leisure spatialities, mainly due to the area occupied by the Cumbuco Village complex (in this case the Portuguese business group Vila Galé).

Clementino and Pessoa (2009) have evaluated the socio-spatial transformations of Natal and its Metropolitan Region. For the authors, the processes that were the basis for the first "metropolitan" interactions date back to the 1980s, and their acceleration converged in the following decade. Coastal spatialities best exemplify the contemporary formation of the Rio Grande do Norte metropolis. The coastal areas of the municipalities of Extremoz and Ceará-Mirim (on the eastern coast) complemented by those of Parnamirim and later by Nísia Floresta (South coast) bring together investments in autochthonous *vilegiatura* practices and those related to allochthonous tourism and *vilegiatura*. Ferreira et al. (2009) identify the continuity of the expansion (in all senses) of these activities, intensifying the spatial hierarchy of the municipalities that are part of the metropolitan region (Fig. 2.8). On this coast, the preferred locations of the touristic dynamics are the beaches of Pirangi (Parnamirim), Jenipabu (Extremoz), and Búzios and Tabatinga (Nísia Floresta).

Fig. 2.7   Metropolitan coast of Fortaleza

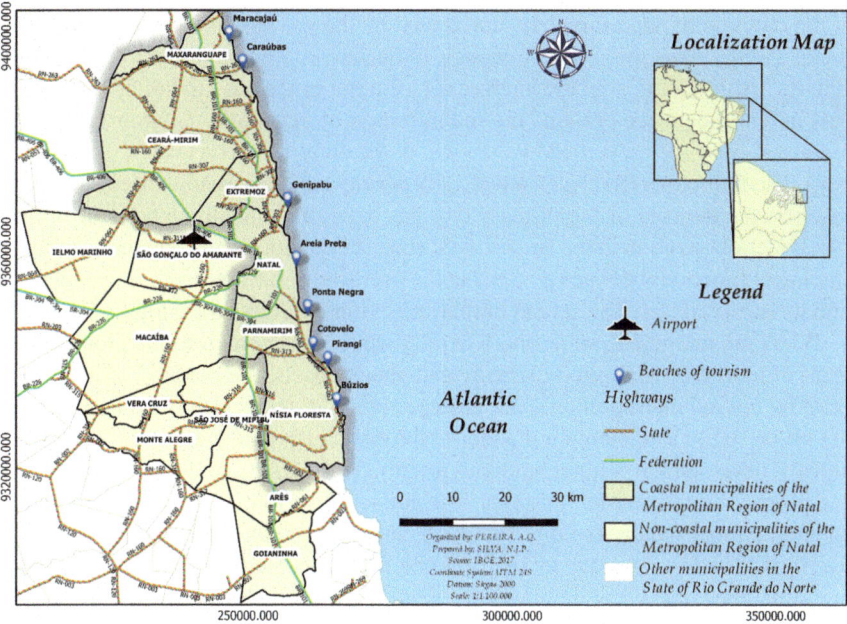

Fig. 2.8   Metropolitan coast of Natal

In the main metropolitan maritime resorts, there is a continuous process of modernization and the inclusion of new leisure practices, notably the central and internationally widespread role of adventure sports, nautical sports, or so-called "nature sports" (Audinet et al. 2017).

2.3 Metropolitan Spatialities of Tourism and Second Homes

All the urban agglomerations or metropolitan regions in the Northeast have a diversity of spatialities that go beyond the municipal political-administrative areas. Internal subspaces form in the metropolises that are not necessarily a continuous urban morphology. With or without planning, the urban nature of modern maritime practices follows and contributes to the formation of some of these spatialities. They are products capable of bringing together different times and subjects. In autochthonous *vilegiatura*, planned (molecular) actions started with real estate developers and even some vilegiaturists, who promoted the consumption of coastal places. Public policies were presented but not in the form of specific plans. The interventions were sporadic and induced by the political activities of both consumers (vilegiaturists) and property developers. During the institutionalization of the MRs, the taste for maritime *vilegiatura* reached sufficient potential to reproduce itself. As it was considered a "social need," this is not surprising. The leisure practices exclusive to the daily life of the elites were approximately reproduced by the growing middle class and other social groups. Was this context enough to declare it a potential metropolitanizing activity? Although it is on the intersection of other processes, and often understood only as a consequence of them, the spatialization of maritime *vilegiatura* efficiently produces subspaces governed by the logic of the emitting urban cluster (the primary city). The four cases (Salvador, Recife, Fortaleza, and Natal) show that in the northeastern case the role of the primary city has been dramatically expanded. In most cases, metropolization in the Brazilian Northeast does not mean complementation or sharing of functions between places, but the overflow of the hub city's social relations into its surroundings. The spread of maritime *vilegiatura* is an example of this.

When the municipal, state, and federal governments perceived the global movement of tourist consumption of sandy and sunny coasts, the planned actions of the State for coastal metropolitan space emerged in a new political-economic context. In the 1990s, regional planning was added to the molecular actions carried out in tandem with the development of autochthonous maritime *vilegiatura*, including the discovery and propagation of coastal locations. The aim was to artificialize space according to an international standard of dissemination, access, circulation, and accommodation, mainly to develop receptive tourism. Would it be possible to wipe clean the slate of maritime *vilegiatura's* spatialities until that point? The touristification of places on the northeastern coast never eclipsed *vilegiatura*, on the contrary, guided by market interests private companies perceived this multiplicity of practices as strategic for the diversification, continuity, and growth of national and international investments. New places were discovered/invented, with a simultaneous redefinition of existing ones, due to the increasing presence of allochthonous vilegiaturists, a fact that in many cases culminated in the annexation of coastal municipalities to the metropolitan perimeter.

Different cases have been recorded worldwide where leisure, tourism, and second homes have restructured the characteristics of entire regions and thus opened up locations to domestic tourist flows, as in South Africa, for example, (Rogerson and Hoogendoorn 2014). There are cases, such as Costa Rica, where the transformations originate from international flows and large-scale tourism and real estate developments (Barrantes-Reynolds 2011). Since the second half of the twentieth century, the Spanish coast (García-Ayllón 2015) and the Australian Gold Coast (Dedekorkut-Howes and Bosman 2015) have become dense urban spaces dedicated, predominantly, to leisure and tourism practices; they are now globalized spaces. These changes have also created numerous natural and socio-economic problems.

The insertion of the metropolitan coast in the Northeast as a place of allochthonous *vilegiatura* has re-dimensioned urban hierarchies, attracting international relationships with the region's metropolises without going through hierarchically superior metropolises (São Paulo and Rio de Janeiro).

The metropolises are very complex. As the state capitals gained relevance, the other municipalities fragmented into coastal spatialities ruled by verticalities (Santos 1996). This fragmentation shows the new contemporary hierarchical patterns, reinforcing the metropolitan model.

Modern maritime practices contribute to metropolitan integration in the Northeast, bringing together two conditions. The first concerns the characteristics of the region's natural processes (sandy beaches, dunes, cliffs, warm seawater, low thermal amplitude with high minimum temperatures), the second shows the continuous planning that reproduces positive images of locations and optimizes or builds the material conditions (densification of technical objects) required to adapt these places to the international consumption model. Thus, maritime *vilegiatura* locations are reproduced amid the following contradictions: city/not city, imitation/distinctiveness, *vilegiatura*/tourism, nature/technical objects, public/private, and planned/contingent.

The metropolitan territorial boundaries and their articulations with the different spatial scales are redefined daily by the diachronic-synchronic interactions of social processes and subjects. The contemporary spatial configuration cannot be explained without revealing the characteristics of the social subjects responsible for the first "race to the sea" as a function of *vilegiatura*. If some affluent families built their second homes in remote seaside villages, the middle class followed them and determined the initial constitution of the peri-urban leisure space in the northeastern metropolises. Northeastern city dwellers envisioned their place in the sun, or rather, on the beach.

Following different methodological paths, the urban forms related to maritime tourism are described by recent case studies: in Pernambuco, Itamaracá (Assis 2001) and Ipojuca (Anjos 2006); in Bahia, the north coast (Mello e Silva et al. Mello e Silva 2009); in Rio Grande do Norte, Parnamirim and Nísia Floresta (Silva 2010); and in Ceará, Aquiraz (Pereira 2006). The production of multiple forms can be divided into three timeframes: until the early 1970s, from the 1970s to the 1980s, and post-1990s.

From the 1960s to the 1970s, it was possible to count the first wave of vilegiaturists and their second homes on the metropolitan coast. The gatekeepers, residents of the "City" in the Northeast, were big businessmen, senior civil servants, and the ruling classes. On the whole, they used their social influence and financial power to secure the purchase (or possession) of land by the sea and built villas, initially selecting the territories of traditional communities. The small straw-and-adobe houses of the fishermen were situated some distance from the sea, even having their backs to the seashore, leaving the beach for anchoring their fishing vessels. The vilegiaturists bought the "plots" directly from the residents of the selected traditional maritime communities, favoring those closest to the Atlantic. Another possibility was the requisition of the right of use from the Secretariat of Patrimony of the Union for the subsequent construction of second residences. Oral reports from many traditional residents complain about the use of illegal instruments by *vacanciers*, who often used land grabbing and the falsification of legal real estate registrations to build their residences.

Morphologically speaking, this first movement involved the construction of horizontal single-family units side by side, forming a line parallel to the sea.

In the City and among their peers, the gatekeepers propagated the benefits of *vilegiatura* and the places they had chosen. This dissemination was enormously persuasive. Among visitors and vilegiaturists, legends spread about the presence of television and sports personalities in specific beach locations. The state itself played a direct role in this movement, by promoting the construction of so-called official vacation homes for state governors on present-day metropolitan beaches.

Many vilegiaturists took advantage of this growing demand and became real estate promoters, amassing lots, and reselling them to those who aspired to *vilegiatura*. These stages coincide with the large-scale expansion of paved roads and electrification networks unrolled in the 1970s, which facilitated the enjoyment of city amenities in the newly discovered *vilegiatura* locations.

From the early 1970s to the late 1980s, real estate companies produced large numbers of urban lots along the metropolitan and non-metropolitan littoral. In the Northeast, this was the first real estate business product made available for maritime *vilegiatura*. The construction of the houses started from a molecular process, or rather an individual process, using mostly local labor, without homogenous construction standards. Thus, in addition to the fishermen's nuclei, second homes began to occupy areas initially designed for modern maritimity, or rather for *vilegiatura*. This period is known as the moment of massification, which included the construction of multi-family real estate developments (condominiums). Until the present, these complexes have formed a significant percentage of the housing stock of second homes. Urban land lots also played a role in opening up "frontiers." The land was fragmented into plots and arranged according to the mercantile-urban pattern, serving as a prototype for larger real estate developments.

Like the pioneering secondary residences and the urban land lots, the construction of the headquarters of leisure clubs belonging to a wide range of professional associations also acted as an attraction. The social constitution of these closed and collective ventures promoted the visits and sojourns in the beach towns by homogeneous socio-cultural strata. Until the 1960s, they had only been located in the hub city, but during the following decades, they spread in a great number to almost all the beaches along the metropolitan coast, imposing themselves as mediators of the massification of users and the constructed forms.

From the 1990s, all the beach towns in the metropolitan municipalities were known to visitors and were home to second residences. The geometric pattern of the localities was driven by the mixed-street model (with diverse areas, associated with random patterns), and in many cases, the dictates of the demand for lots and second homes predominated. The municipalities played a supporting role in the process: firstly, by supporting the dynamics of the real estate market, and afterward, by building basic infrastructure, especially the paving of the streets and access roads.

From the residues of the previous decades (regarding the relationship between the permanent housing stock and the second residences), two groupings were formed: homogeneous and heterogeneous densities. The former are primarily second residences occupied by vilegiaturists and composed of single-family units and horizontal and vertical condominiums. The second older group is a complex mix of residents and vilegiaturists, with permanent and occasional residences that are conventionally formed by single-family dwellings. In most cases, the homogeneous clusters are the expansion of the heterogeneous ones, forming an urban spot that grows parallel to the sea. However, this does not mean that the whole metropolitan coastline is a continuous agglomeration.

There are still spaces of unbuilt fields and areas that host other productive and (re)productive activities. Over the last three decades, it is noteworthy that improved logistical conditions have consolidated the process of metropolitan inclusion of these centers. This improvement is understood in terms of the possibilities offered to the segments living in the primary cities, which also logically includes those who live at the other extremity of the network: coastal clusters.

In the twenty-first century, federal public resources and resources from the federal units have been used to build international airports and highways parallel to the coast. National and international groups have started a process of articulation with local owners and investors. Whether calculated or not, the state governments validate the expansion of the process by continuing to implement urban infrastructures, mediating links between local, regional, and international companies, publicizing the "vocation" of the locations, and granting environmental licenses for the construction of projects. The latter is responsible for two sets of changes in the production of metropolitan coastal space.

Firstly, the construction of new typologies (tourist-hotel complexes, resorts, cond-hotels, flats, and condominiums) prioritizes the space surrounding the city hub, encouraging the state to complement the technification process of metropolitan coastal space (transportation, circulation, and supply).

Secondly, most of the ventures are not located in places consolidated by generic tourist activities and may be discontinuous, thus forming self-sufficient closed spaces (offering everything necessary for the stay, including access to the beach). In terms of urban morphology, these ventures promote synergistic effects: increasing the number of plots sold in previously divided urban subdivisions, building new urban subdivisions, and building smaller projects (condominiums and flats). They also increase the number of single-family second homes. The dynamics feed back and enable the confluence of new and old patterns of land use.

One of the indicators of these transformations is the participation of foreign groups in the acquisition of land and the construction of tourist-real estate ventures. Between 2001 and 2007, the states of Bahia, Ceará, and Rio Grande do Norte received significant volumes of foreign capital in the sectors of real estate development and construction of tourist-residential typologies. With the 2008 economic crisis, foreign resources declined, and projects were redefined or canceled. The metropolitan area of Natal is a clear example of these redefinitions, due to the number of canceled or delayed projects.

The consolidation of major projects occurred in the metropolitan areas of Fortaleza, Salvador, and Recife. Examples are the complexes of Hotel Costa do Sauipe (Mata de São João-MRS), Aquiraz Riviera (Aquiraz-MRF), and Paiva Reserve (Cabo de Santo Agostinho-MRR). The locations are not random and can be explained by the economic, infrastructural, demographic, and touristic importance of these metropolitan regions and their resorts.

References

Almeida PH (2006) A economia de Salvador e a formação de sua região metropolitana. In: Carvalho IMM, Pereira GC (eds) Como anda Salvador, 2a edn. Edufba, Salvador, pp 1–50
Alves PRM (2009) Valores do Recife: o valor do solo na evolução da cidade. Luci Artes Gráficas, Recife
Anjos KL (2006) Turismo em cidades litorâneas e seus impactos ambientais urbanos: o caso de Porto de Galinhas, PE. Dissertação de Mestrado – UFPE. Programa de Pós-graduação em Desenvolvimento Urbano. Recife, PE, BR
Assis LF (2001) A difusão do turismo de segunda residência nas paisagens insulares: um estudo sobre o litoral sul da Ilha de Itamaracá-PE. Dissertação de Mestrado. Centro de Filosofia e Ciências Humanas – UFPE. Recife, PE, BR
Audinet L, Guibert C, Sebileau A (2017) Les "Sports de Nature": une catégorie de l'action politique en question. Édition du Croquant, Paris
Barrantes-Reynolds MP (2011) The expansion of "real estate tourism" in coastal areas: its behaviour and implications. Recreat Soc Afr Asia Lat Am 2(1):51–70
Bezerra JS, Araújo LM (2007) Reestruturação e centralidade: breves notas sobre a cidade de João Pessoa. URBANA (CIEC/UNICAMP), ano 2, n° 2. Dossiê: Cidade, Imagem, História e Interdisciplinaridade, pp 1–16
Carvalho IMM et al (2004) Polarização e segregação socioespacial em uma metrópole periférica. CADERNO CRH 17(41):281–297, Mai./Ago
Clementino MLM, Souza MAA (orgs) (2009) Como anda Natal e Recife. Letra Capital/Observatório das Metrópoles, Rio de Janeiro
Clementino MLM, Pessoa ZS (orgs) (2009) Natal: uma metrópole em formação. Natal: Observatório das Metrópoles. EDUC, São Paulo
Correa SMS (2010) Germanidade e banhos medicinais nos primórdios dos balneários no Rio Grande do Sul. História, ciência, saúde – Manguinhos 17(1):165–184

Costa MC (2006) Lustosa. Clima e salubridade na construção imaginária do Ceará. In: Silva JB et al (eds) Litoral e Sertão: natureza e sociedade no nordeste brasileiro. Expressão Gráfica, Fortaleza, pp 73–82

Costa MC (2007) Fortaleza: expansão urbana e organização do espaço. In: Silva JB et al (orgs.). Ceará: um novo olhar geográfico, 2ª edn. Edições Demócrito Rocha, Fortaleza, pp 51–100

Dantas EWC (2002) Mar à Vista: estudo da maritimidade em Fortaleza: Fortaleza: Museu do Ceará, Secretaria de Cultura e Desporto

Dantas EWC (2006) Cidades litorâneas marítimas tropicais: construção da segunda metade do século XX, fato no século XXI. In: Silva JB et al (eds) Panorama da geografia brasileira 2. Annablume, São Paulo, pp 79–89

Dantas EWC (2010) Antecedentes do turismo no Nordeste. In: Dantas EWC et al (eds) Turismo e imobiliário nas metrópoles. Rio de Janeiro: Letra Capital, pp 17–34

Dantas EWC et al (2010) Turismo e imobiliário nas metrópoles. Letra Capital, Rio de Janeiro

Davidovich F (2004) A "volta da metrópole" no Brasil: referências para a gestão territorial. In: Ribeiro LCdeQ (org) Metrópoles: entre a coesão e a fragmentação, a cooperação e o conflito, São Paulo, Fundação Perseu Ábramo; Rio de Janeiro: FASE, pp 197–229

Debié F (1993) Une forme urbaine du premier age touristique: les promenades littorales. MappeMonde. 1:32–37

Dedekorkut-Howes A, Bosman C (2015) The gold coast: Australia's playground? Cities 42:70–84

Ferreira AL et al (2009) Dinâmica imobiliária, turismo e meio ambiente: novos cenários metropolitanos. In: Clementino MLM, Pessoa ZS (orgs) Natal: uma metrópole em formação. Natal: Observatório das Metrópoles. EDUC, São Paulo

García-Ayllón S (2015) La Manga case study: consequences from short-term urban planning mass destiny of the Spanish Mediterranean coast. Cities 43:141–151

Lefebvre H (2004) O direito à cidade. Tradução de Rubens Eduardo Frias, 3ª edn. Centauro, São Paulo

Linhares P (1992) Cidade de água e sal. Por uma antropologia do Litoral Nordeste sem cana e sem açúcar. Fundação Demócrito Rocha, Fortaleza

Mello e Silva SB et al (2009) Globalização, turismo e residência secundária: o exemplo de Salvador-Bahia e de sua região de influência. Observatório de Inovação do Turismo – Revista Acadêmica IV(3):15pp

Miranda LIB (2008) Organização socioespacial e mobilidade residencial na Região Metropolitana do Recife, PE. Cadernos Metrópole, N. 12:123–144

Pereira AQ (2006) Veraneio marítimo e expansão metropolitana no Ceará: Fortaleza em Aquiraz.. Dissertação de Mestrado – UFC. Programa de Pós-graduação em Geografia. Fortaleza, CE, BR

Rogerson C, Hoogendoorn G (2014) VFR travel and second home tourism: the missing link? The case of South Africa. Tour Rev Int. 18(3). https://doi.org/10.3727/154427214x14101901317156

Santos M (1996) Metamorfoses do espaço habitado. Fundamentos teóricos e metodológicos da geografia, 4ª edn. HUCITEC, São Paulo

Silva KO (2010) A Residência secundária e o uso do espaço público no litoral oriental potiguar. Dissertação de Mestrado, Departamento de Geografia, Universidade Federal do Rio Grande do Norte, Natal, RN, BR

Silva HRF et al (2009) Dinâmicas Metropolitanas de Salvador: um estudo dos municípios de Lauro de Freitas, Camaçari e Mata de São João. In: Anais do XI Simpósio Nacional de Geografia Urbana, pp 1–20

Sousa AN (2010) Orla oceânica de Salvador: um mar de representações. Dissertação de Mestrado – Instituto de Geociências, Universidade Federal da Bahia, Salvador, BA, BR

Souza MS (2006) Segregação socioespacial de Fortaleza. In: Silva JB et al (orgs). Litoral e Sertão: natureza e sociedade no nordeste brasileiro. Expressão Gráfica, Fortaleza, pp 149–162

Vasconcelos PA (2002) Salvador: transformações e permanências (1549–1999). Editus, Ilhéus

Villaça F (2001) Espaço intra-urbano no Brasil, 2ª edn. Studio Nobel, Lincoln Institute, São Paulo

Chapter 3
Coastal Metropolitization and Tourist-Real Estate

Abstract The preference for and the exercise of maritime practices give rise to diverse structures and result in leisure-focused urban spaces in the metropolitan context. This chapter presents the Northeast coast of Brazil as a case study; however, connections are also made with other national and international spaces. The starting point is the empirical-conceptual construction of the process used in the analysis of these dimensions: urban morphology and social practices.

Keywords Brazil · Urban morphology · Coastal village

3.1 Introduction

The process of contemporary urbanization is marked by the interaction between forms and contents with distinct temporal densities. Old and new city functions interact with old and new fixtures, leading to the (re)production of heterogeneous and, in a contradictory but simultaneous way, homogenous urban spaces (under particular scales). Therefore, any study which only explores one dimension of urban functions is increasingly less able to explain the diversity mentioned above.

On discussing theses about urbanization, Brenner (2014) indicates three series of urban spatial transformations since the start of the millennium: (i) the appearance and explosion of new, larger morphologies, which crossover urban and rural spaces; (ii) the many varied state actions aiming to make urban spaces attractive to national, international, and transnational investment; (iii) the transformation of the metropolis into a strategic space for the territorial conflicts between distinct social segments. Both Brenner's first and second items are essential themes in this chapter.

Observations of the urban in the twenty-first century show that the metropolization of space has become a scientific frontier to explore. Consequently, metropolitan spaces unfold through increased density and functional, real estate, and demographic diffusions. The contribution of metropolitan studies touches on the concept of the existence of agents who promote restructuring (market, science, technology) but it also includes creators of specific (and even particular) conditions: inherited urban functions, urban sites, and regional and local political relationships.

© The Author(s), under exclusive license to Springer Nature Switzerland AG 2020   35
A. Queiroz Pereira, *Coastal Resorts and Urbanization in Northeast Brazil*,
SpringerBriefs in Latin American Studies,
https://doi.org/10.1007/978-3-030-46593-3_3

Discussing the characteristics of metropolized spaces, Lencioni (2013) points to the initial evidence provided by the landscape, as "[...] to the extent that one becomes more distant from the spaces where there is a greater density of persons, goods, and flows, non-metropolized spaces impose themselves upon metropolized spaces" (p. 19).

It is important to emphasize the relevance of how on specific scales, preponderant phenomena bring about the process of metropolization of space. This chapter proposes an emphasis on the relationship between urbanization, metropolization, and leisure practices in urban agglomerations located on oceanic coasts. The preference for these practices and the resulting actions give rise to diverse structures and result in metropolitan leisure spaces. The underpinning principle is understanding the journey and the temporary stay based upon leisure as an urban and maritime practice. In current times, *vilegiatura* represents the totality of the social practices ensuing from the temporary stay commonly linked to leisure. Rather than distinguishing between tourists and vilegiaturists, geographical studies from this perspective focus their analyses and understanding of the territorial impacts on the nodes of the "net" promoted by this urban fixation.

Instead of following a methodological path close to that of the geography of tourism, or a new branch (geography of leisure), it is wise to follow an urban geography analysis which emphasizes *vilegiatura*, understood as both bringing about the urbanization process and being brought about by it. The intention is to understand the capacity of this practice to organize locations and establish relationships, inserting them into the composition of metropolitan spaces.

3.2 Maritime Vilegiatura and Its Urban Dimension

The temporary stay, especially in coastal spaces, is a catalyst, which synthesizes and causes a vast diversity of leisure and rest practices. Locations and subspaces are reconfigured, modernized, and urbanized to meet the needs of the crowds moving between places. This happens because the origin, formation, and daily life of the *vacanciers* are in the city and, primarily, the urban. At present, it is the spaces next to the sea that are passing through these transformations most intensely, as they are desired for *vilegiatura* and/or housing around the world.

Among many options, the beach is urban society's preferred space for a temporary stay for leisure purposes. Boyer (2008) considers Brighton, on the English coast, the first modern beach as a result of the enjoyment of the country's elite in the nineteenth century; later, the beach was used by the less noble segments of British society. The organization of leisure complexes on the coast was the golden age in maritime *vilegiatura* in Europe. Undoubtedly, there was a strong urban function in the sociocultural activity of this movement. It became the prototype for the reinvention of urban use on the coastal boundaries at middle latitudes, especially in the Mediterranean and Florida.

This context is directly associated with the ascension of industrialist values in Western civilization, a moment in which urban society became a virtuality (Lefebvre 1999). There is a series of transformations, among them the redefinition of the perception of the beach, transfigured into a social space. The territory of emptiness indicated by Corbin (1989) became a territory for potential urbanization. City dwellers walk on the beach and live by the beach, rebuilding it to meet their leisure and well-being needs. Promenades were built along the beaches of nineteenth-century cities, and seaside walks reigned. By the first quarter of the twentieth-century automobiles had conquered the avenues along the beachfront (Dibié 1993). From then on coastal cities embraced their seaside.

Beaujour-Garnier and Chabot (1963) in his *Traité de géographie urbaine* recognizes the character of urban function in leisure. The author describes various types, giving the *villes de bains de mer* as fabulous examples of urbanization based upon the taste for seasonality and leisure. If the process of a relationship between urbanization and maritime *vilegiatura* is evident, by the action of the first upon the second, other authors, specifically Pereira (2014), have identified the practice of leisure as a potential cause of the dissemination of urban forms and contents. There is, therefore, a resemblance to the characterization of the French author in the 1960s.

> Ces villes de récréation relaient em quelque sorte les autres formes de vie urbaine; elles sont donc particulièrement nombreuses dans les pays où le réseau urbain est dense, les agglomérations fortes et le niveau de vie élevé (Beaujour-Garnier and Chabot 1963, p. 178).

The locations linked to these urban populations' leisure activities reorganize themselves continuously so that the principal functions and sceneries become clear: second homes, hotels, restaurants, condominiums, resorts, marinas, shops, and services. During the twentieth century, the variation in the concentration of these elements formed distinct coastal sceneries redefined by the desire of city populations for leisure. This process became a "tributary" of the general process of urbanization, due both to the emergence of *vacanciers* (originating from medium-sized and large urban agglomerations) and the social-spatial transformations in the development of health resorts. In Lefebvrian terms, an extension occurs in the urban fabric.

In the process of increasing the value of coastal spaces, urban characteristics are added to the scenery following the same rhythm as the incremental growth in the flow of people and investments. The initial production of the locations follows different paths, which then converge in the heterogeneity of subjects that is also a hallmark of the mass standardization of the process. In this sense, there is the participation of individual real estate entrepreneurs (*gatekeepers*), business groups, and even trail brazing users disinterested in monetary gain. After a particular point, they act simultaneously, with or without a conflict of interests. With the institution of maritime *vilegiatura*, locations go through transformations which are indicative of urbanization, namely:

(a) **New property dynamics**: Distinct uses are added to properties, with an increase in their value and the creation of a real estate market based on exchange value. The concept of rarity has a direct impact, leading to a relative increase in prices. The division of land appears and expands, generally regulated by legislation specific to urban spaces. At the same time, residential constructions multiply in varied forms: residences for seasonal use by a single-family, residential units in horizontal and vertical condominiums, and tourist property complexes (which take the form of a planned resort). All these elements contribute to the implementation and/or densification of the urban morphology in the locations.

(b) **Diversification in the urban division of labor**: Transformed into economic activity, beyond physical infrastructure these forms of leisure require a range of services and shops. In this sense, there is a progressive dislocation of labor posts toward the tertiary sector at locations by the sea. This is due to both the arrival of experienced business groups and the formation of local entrepreneurs. The seasonality of *vacanciers* interferes with the quantitative distribution of jobs and the flexibility of seasonal functions. Public and private organizations train professionals using technical training resources based on external and international standards of service provision.

(c) **Positive demographic dynamic**: Due to the insertion of new economic activities and the creation of jobs, these spaces become receptive to migratory flows. A common phenomenon for activities marked by spatial mobility such as *vilegiatura* and tourism is the transformation of *vacanciers* into residents. In this way, beyond vegetative and migratory growth, such spaces display quantitative population growth.

(d) **Cultural contacts and establishing customs**: The social interaction between *vacanciers* and those living in the receiving spaces promotes symbolic and cultural exchanges, as well as commercial relations. Urban fashions and customs are adopted in these locations. These influences are notable in terms of clothing, the use of electronic equipment, the architecture of homes, leisure activities, and even every day verbal expressions.

(e) **Urban legal institutions**: Due to the aspects above, coastal spaces are invariably regulated by local legislation that considers them as urban zones and/or areas. Laws concerning zones, land use and occupancy, construction codes, and postures are examples of legal instruments regulating access to these spaces and their transformation. At the same time, a structure is created for collecting taxes, especially those on the ownership of properties and the market transfer of goods. This legislation is an essential condition for the spread of the urban fabric, as it indicates priority areas for expansion. On the other hand, due to natural characteristics, environmental protection legislation has limited land use across the world, and its main justification is the vulnerability to erosive processes and the need to conserve native flora and fauna.

Highlighting these effects demonstrates the role of *vilegiatura* in expanding the urban fabric. As a result, the maritime leisure *villas* move from isolation to integration in a complex whole: the metropolis.

3.3 Northeast of Brazil

In the twenty-first century, the metropolization of space corresponds to the formation of urban agglomerations that are not necessarily continuous in terms of space, and with distinct levels of integration. This integration is evidenced by the density of the infrastructure (production, transport, and energy, among others) generally articulated in networks. This territorial whole is distributed heterogeneously. The contemporary metropolis is a complex of functions, that is, subspaces.

As argued above, the formation of metropolises occurs through the conjugation of metropolized and non-metropolized spaces (Lencioni 2013). However, the most crucial question addresses the issue of which are the most and least efficient vectors and/or social activities in the current production of spaces and metropolitan networks. In this sense, there is little doubt concerning the roles played by the diffusion of different industrial sectors and, traditionally, that of residential properties, associated with the creation of large-scale commercial centers (mainly outlets and shopping malls). Notwithstanding the desire for leisure on the part of the masses and the elites, there is a strong potential to extend the urban fabric and generate an interconnection between territories. In this way, the traditional city no longer meets all the desires and possibilities regarding leisure.

The process of metropolization and the metropolis itself opens up a range of options. The distinct places are included as points in a network, destined for temporary leisure stays. Analyzed by Beaujeu-Garnier in the 1960s, this context is not exclusive to the current times.

> Chaque grande agglomération a ainsi ses annexes villégiatures où la famille passe ses longues vacances, où le chef de famille se rend pour le week-end, et c'est surtout vrai quand on est proche de la côte, où les plages deviennent facilement des petites villes (Beaujour-Garnier and Chabot 1963, p. 432).

Rather than being an exception, the urbanization of spaces for leisure functions (and for *vilegiatura*) is typical in the Canadian, Irish, Australian, New Zealander, South African, Spanish, and Scandinavian cases (Hall and Müller 2004). In many situations, this process is understood as strategic for the development of the receiving regions, mainly motivated by tourism and the concentration of second homes, for both vacation and weekends. Specifically for the coast, studies by Roca et al. (2009), Latorre (1989) discuss the influence of large cities in the formation of new urbanized areas in Portugal and Spain, respectively.

Once again, there are two variables in the process: a social variable and a spatial one. The first variable refers to the mass increase in demands for practices near to *vilegiatura* and coastal tourism (Pereira 2014); the second refers to the urbanization of "underdeveloped" countries during the second half of the twentieth century. The demographic growth associated with a diversification of the productive base (via state interventions—public policies) conditioned the formation of metropolises in subnational regions. In this context, the trend in the Western World is for the gradual incorporation of coastal spaces adjacent to the metropolises.

In the Brazilian case, with a few exceptions, coastal location is a relevant factor in the formation of the main national metropolises. In this process, the expansion of leisure zones on the coast also has an intense expression, especially those located in the states of Rio de Janeiro and São Paulo. In the first case, the state capital is recognized for its inherent coastal nature. In the twentieth century, the image of the archetypal Rio de Janeiro resident was formed, usually as a lover of the sea and the maritime. There are direct impacts, both on the urban beaches (among the most well known in the world) and on the other coastal spaces in the state of Rio de Janeiro. In the case of the state of São Paulo, the area surrounding the port of Santos had strong links to the demands of the metropolis, becoming the densest urban "outskirts" produced for maritime leisure in the country. On the main extended holidays, including the New Year celebrations, hundreds of thousands of residents of the city of São Paulo travel down the mountain range to the coast, toward the beaches of Santos, Granjá, Praia Grande, São Vicente, and Bertioga, among others.

The northeast coast of Brazil has three important metropolitan regions (Salvador, Recife, and Fortaleza) and is a byword for the national tourist coast. From the 1990s, the image of a tropical paradise, with sandy, sunny beaches, and warm waters, has been consolidated, permitting the insertion of maritime resorts as new agents in coastal spaces previously restricted to local autochthons. Currently, *vacanciers* for *vilegiatura* from other regions and other countries (non-native) have joined traditional flows. These flows help create the notion of tourist metropolization.

Maritime *vilegiatura* is reproduced as it creates important peri-urban areas related to large coastal agglomerations, especially metropolises. This specific peri-urbanization defines a characteristic form of metropolization which unfolds in normative, socio-economic, and real estate transformations in coastal spaces. It is essential to indicate theoretical and methodological elements capable of explaining this process in the Northeast of Brazil. The starting point is the empirical and conceptual construction of the process which establishes two analytical dimensions: urban morphology (norms—urbanism—real estate) and social practices. Empirical evidence, the primary social agents, and the key concepts are the principal tools to explain the content of these dimensions.

3.4 Urban Morphology: Property, Urbanism, and Norms

Urban morphology is a key indicator of socio-territorial transformations in coastal spaces. It happens in a process linked to property dynamics, regulated by urban planning that uses urbanistic technical knowledge. In the case of *vilegiatura* in the metropolis, the demand for maritime leisure practices organizes the territory through the installation of a mobility infrastructure (coastal highways) and, principally, in a set of services and properties which provide the conditions for a temporary stay.

Urban morphology is the materialization of the network. The central point is the city, the origin of the metropolitan region, and the process of coastal peri-urbanization, which is spreading across the world, presenting general characteristics as indicated by Terán (1969).

La urbanización de las zonas periféricas y de los ambientes rurales circundante se xtiende for-mas de vida urbana, sin que ileguen a crearse estructuras urbanas. Vastos espaciosinorgánicos se incorporan a la urbe, aboliendo das fronteras entre lo urbano y lo rural. Un 'habitat' de tipo urbano se dispersa y desparrama sobre território aún no urbanizado, dando lugar a essas zonas de calificacióndudosa: 'suburbanas', 'interurbanas', 'exurbanas', 'rurbanas', etc., en las que se pierde el concepto tradicional de ciudad, la cual se haceasí difícilmente abarcable y comprensible em forma y dimensión por los habitantes. Aquarone ha explicado así la for-mación de estes nuevo ente semiurbano, extendido ampliamente, que puede englobarem su trama várioscentrosantiguosmás o menos importantes y más o menosconcentrados (Terán 1969, p. 130).

In the case of the Northeast, peri-urbanization is the expression that explains the incorporation of the beach areas into the urban, or metropolitan dynamics. Territo-rially, the formation of nuclei of occupation and the insertion of technical systems into the territory results in a multifaceted, disfigured urban morphology (Lencioni 2013). The touristic metropolization of the Northeast, explained by Dantas (2015), is yet another indicator of the socio-spatial diversity of this region at the beginning of this century. In terms of planned and moderated actions, the metropolization of the coast has a linear and fragmented urban morphology, made more dynamic by seasonal movements.

Among the explanations for this linearity and fragmentation are the natural loca-tion, the road transport matrix (which runs parallel to the coastline), and the spon-taneity of the formation of the coastal locations. The seasonality is explained by the regulation of the use of time and space in the urban lifestyle, which separates specific periods and spaces for the development of certain practices, among them leisure. This is fundamentally a consequence of islands of occupation, considered fragments of the metropolis, as shown in the model in the figures below (Figs. 3.1 and 3.2).

In all the urban-metropolitan agglomerations in the Northeast, spatiality goes beyond the municipal polygons. The territories produced for leisure form a network which is internal to the metropolises and has a discontinuous urban morphology; nonetheless, they are integrated through airports and coastal roads. Although it is at the interstice of other processes and often understood solely as a consequence, the spatial occupation of *vilegiatura* is efficient in the production of subregions regulated by the logic of the emitting urban agglomeration.

This fragmentation and linearity can usually be explained by the increase in the value of the land closest to the sea and the invention of new places suitable for recreational activities. Therefore, in the peri-urban space, a mosaic is formed, con-sisting of resorts, environmental protection areas, uncultivated estates, small fishing, and extractive villages, as well as illegal and pauperized settlements. Thus, there is heterogeneity in the spaces that are being metropolized.

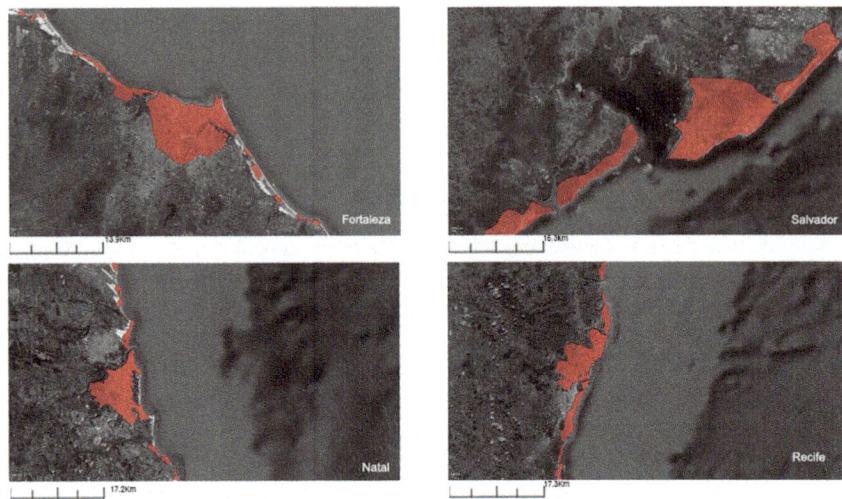

Fig. 3.1 The urban coastal metropolitan sprawl in Fortaleza, Natal, Recife, and Salvador. *Source* Google Earth Pro. (2015)

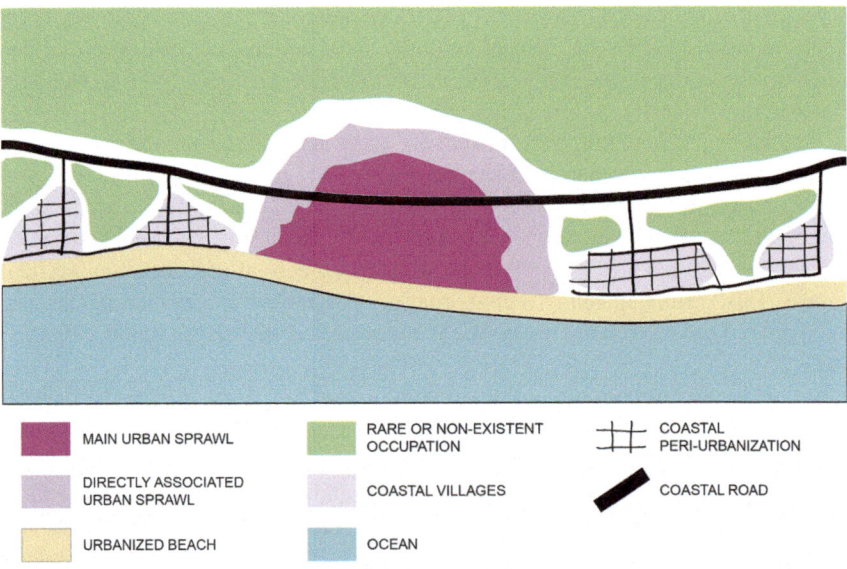

Fig. 3.2 Model of the occupation of the coastal metropolitan space in the Northeast. *Source* Created by the author

For each metropolitan unit, a central hub and a set of articulated maritime leisure resorts are formed. A network is established, consolidated by centrifugal and centripetal movements. The hub city attracts visitor flows and has the brand, the attractive positive image, the hotels, the rental flats, and the exciting nightlife. It has globalized specialized services. The peri-urban space has the beach, a lower population density, environmental quality, and the resort. The hub city has company headquarters, planning, and resources, while the products, apartments, homes, and land units are on the beach. The hub city and the resort nodes form the network of places destined for leisure and become increasingly evident as the emblem of the metropolises of the Northeast.

In the 1990s, the maritime *vilegiatura* nodes of the metropolitan network were formed through the construction of second homes, starting in small villages with traditional communities (mainly fishing). In the 2000s, international demand and the offer of new real estate products facilitated the installation of tourism and property complexes. These are mixed-use megaprojects, characterized by a diversity of activities, including the letting and sale of properties destined for maritime leisure. They are capable of attracting national and international flows of holidaymakers, consuming areas of more than one hundred hectares. Dozens of such enterprises are concentrated in the metropolitan areas of the Northeast and are responsible for two sets of changes in the production of metropolitan coastal space. Firstly, the construction of new typologies (resorts, hotel condos, flats, and condominiums) prioritizes the space surrounding the hub city, leading the State to complement the process of upgrading the infrastructure in the coastal metropolitan space (mainly the coastal transport routes). Secondly, the location of these enterprises does not necessarily occur in places consolidated by generic tourist activities; they may be discontinuous and thus form self-sufficient closed spaces (offering all that is necessary for the tourist's stay). Due to public investment and the development of tourist property complexes space which was formerly regulated by spontaneous uses becomes induced use. The magnitude of flows, spatial transformations, the redefinition of places, and the formation of new territorial networks increases (Figs. 3.3, 3.4, and 3.5).

One notable transformation is the significant increase in new property developments, especially those for occasional use (IBGE), which are very similar to those considered second homes. According to data from the last censuses, there has been an increase in the number of such homes, especially since the 1990s. The states with the most second homes also have the metropolises with the highest number of this type of property, namely, Salvador, Recife e Fortaleza (Table 3.1).

It is believed that in the next census counts, and with the construction of complexes and their internal enterprises (condominiums, land shares), the growth in the number of second homes will continue.

Unlike the diversity of public and private investments, the coastal sectors coordinated by the metropolis display the characteristics of precarious metropolitan urbanization in their internal urban scale. Socio-spatial problems linked to the massive growth of maritime leisure have emerged, associated with the spread of *vilegiatura* and its property derivatives. The main adverse effects related to this are the excessive subdivision of the coastal ecosystems and the constant increase in land prices, even

Fig. 3.3 Mosaic describing urbanistic and architectural forms of tourist property enterprises on the metropolitan coast of Fortaleza (2015). *Source* Google Earth Pro and fieldwork

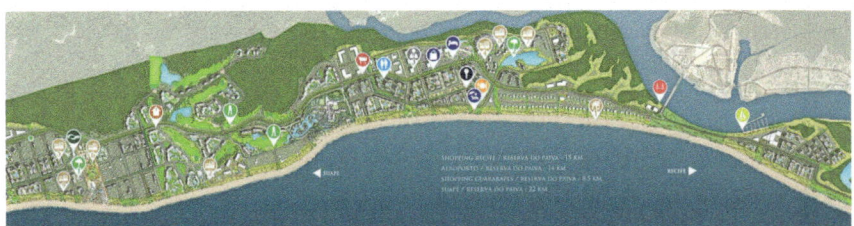

Fig. 3.4 Master plan of the tourist property complex Reserva do Paiva. Cabo de Santo Agostinho—Recife. *Source* http://www.reservadopaiva.com.br/img/mapa.jpg

leading to the emergence of deprived and irregular neighborhoods. The common contradictions in the formation of Latin American cities are also reproduced on the beach.

The impacts caused by the construction and operation of major tourism and real estate projects, such as those built in Fortaleza, Recife, and Salvador, are complex

Fig. 3.5 Master plan of the tourist property complex Iberostar. Mata de São João –Salvador. *Source* http://slideplayer.com.br/slide/3688908/

Table 3.1 Homes with occasional use in the metropolitan regions of Fortaleza, Natal, Salvador, and Recife

Metropolitan region	Census			
	1980	1991	2000	2010
Fortaleza–Ceará	4983	15,530	26,564	39,139
Recife–Pernambuco	6807	21,968	31,321	45,185
Salvador–Bahia	10,136	30,545	46,102	70,502
Natal–Rio Grande do Norte	2497	6995	12,963	25,238

Source Summary of IBGE Census 1980, 1991, 2000, 2010

and generate a wide range of conflicts. These problems are not an exception in the international scenario. Burt (2014) studied major projects on the Persian Gulf coast of the United Arab Emirates and highlights the environmental impacts on reefs and mangroves. Barrantes-Reynolds (2011) recollects the conflicts in Costa Rica due to the implementation of tourism and real estate projects and the local population's use of water resources. Rutty and Scott (2014) warn of temperature-related climate changes, thermal comfort, and their relationship with the perception of tourists and their leisure practices. When it comes to urbanism, the law, urban planning, and public-private relations, the problems are not fewer or simpler. There is difficulty in legislating and regulating uses (Persson 2015), prioritizing public investments and avoiding social injustices (Akyol and Cigdem 2016). Likewise, organizing the real estate market and, especially, the growth of agglomerations (Garcia-Ayllón 2015; Dedekorkut-Howes and Bosman 2015) poses challenges. At an urbanistic level, there are fractures in the urban fabric that occur because the technical quality of the plans is restricted to the interior of complexes, meeting the "urban" demands for comfort, social ostentation, and the sensation of security. The peri-urban surroundings are not taken into account, and the landscaping is varied, due in part to new building standards that diverge from previous designs, and also due to new walls and barriers which do not allow direct contact with previously existing settlements (second homes following vernacular standards and fishing communities, among others). The limited access to spaces on the seafront and the poor maintenance of the transport routes which allow circulation are reliable indicators of the dislocation of the urbanistic approach centered on private enterprise and not on the location as a whole (Fig. 3.6).

As already mentioned, the integrated tourist complexes, with their gated, privately regulated environments were the novelties of the Northeast's coastal space at the beginning of the twenty-first century. However, almost two decades later, the main projects have not been completed, mainly due to the international and national economic crises (Fig. 3.7). One of the greatest examples is the number of planned and unexecuted developments in the state of Rio Grande do Norte. Projects in Recife (Pernambuco) and Fortaleza (Ceará) have reviewed certain plans, such as the construction of golf courses. In the Reserva do Paiva complex, the proposed course has not been built and the one in the Aquiraz Riviera complex is underutilized. In other cases, real estate-tourism projects have changed their focus from residential tourism to the sale of units to the local primary residence market. These conditions have marked the heterogeneity of the occupation of metropolitan coastal spaces in Brazil.

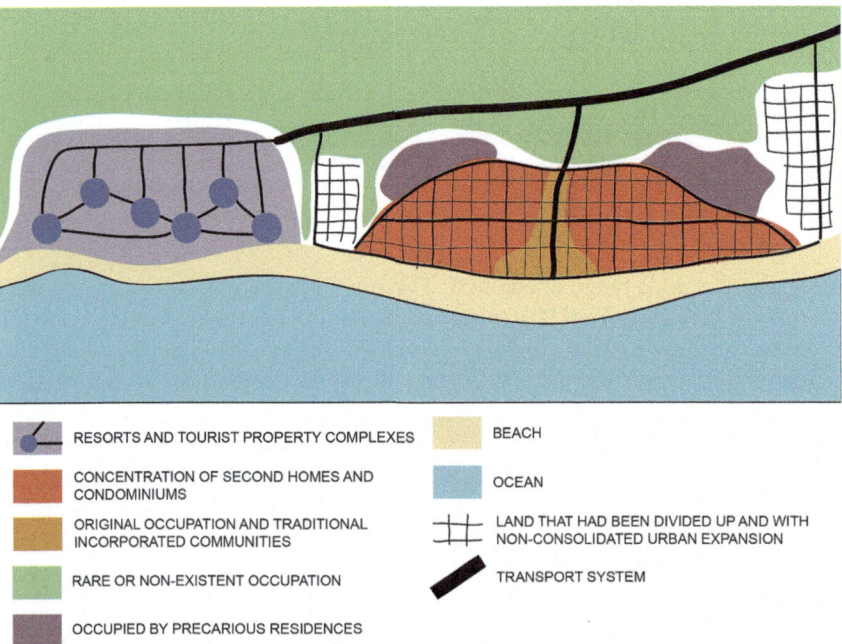

	RESORTS AND TOURIST PROPERTY COMPLEXES		BEACH
	CONCENTRATION OF SECOND HOMES AND CONDOMINIUMS		OCEAN
	ORIGINAL OCCUPATION AND TRADITIONAL INCORPORATED COMMUNITIES		LAND THAT HAD BEEN DIVIDED UP AND WITH NON-CONSOLIDATED URBAN EXPANSION
	RARE OR NON-EXISTENT OCCUPATION		TRANSPORT SYSTEM
	OCCUPIED BY PRECARIOUS RESIDENCES		

Fig. 3.6 Predominant forms and types in the coastal space of the metropolises of the Northeast. *Source* Developed by the author

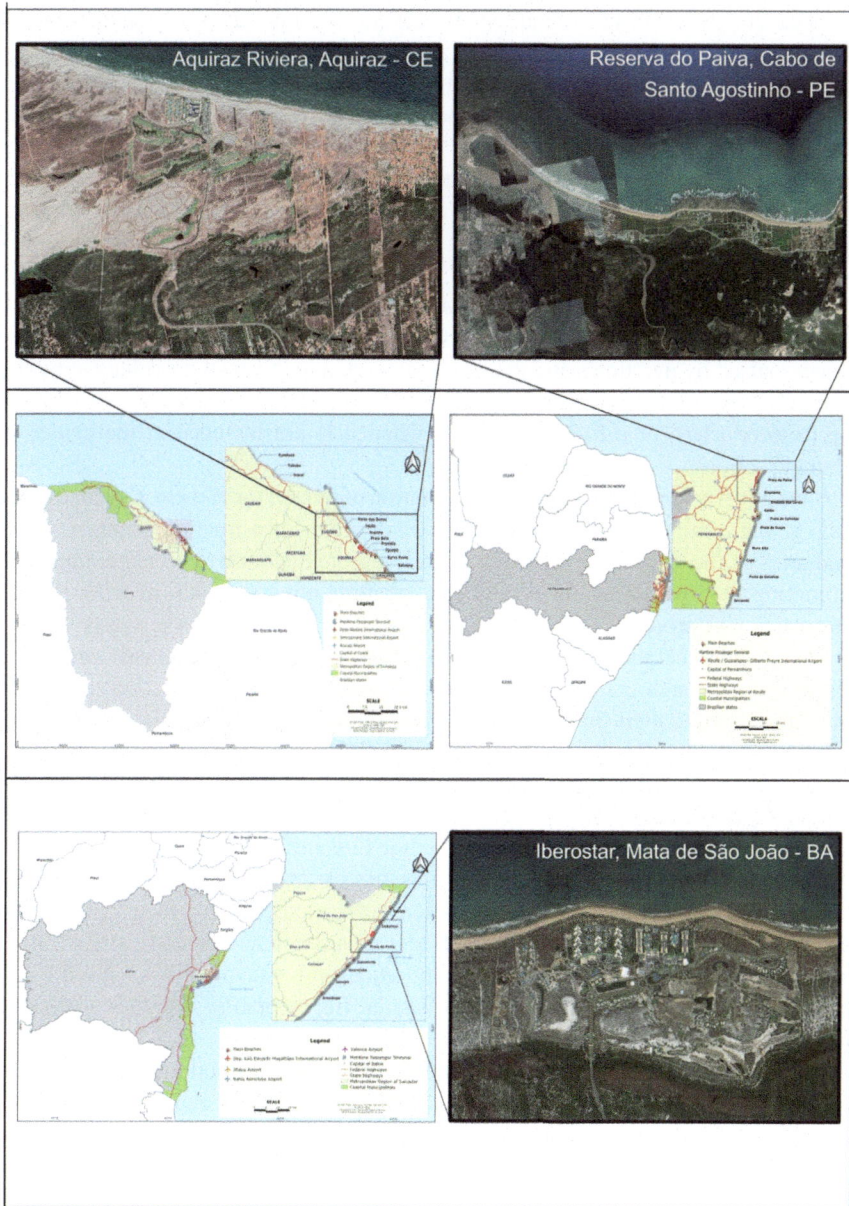

Fig. 3.7 Complex tourist developments in Bahia, Ceará, and Pernambuco, 2020. *Source* Google Earth Pro

The metropolitan beaches, nodes in the maritime leisure network in the metropolis, still have the stigma of a duality of representations: attractive and repulsive. This movement has an incremental impact on the dynamic of the transformations of peri-urban coastal space. Fragments are produced in the urban fabric, differentiated in terms of space and time, according to the social strata which they attract and/or repulse. At times these are "urban-natural" planned paradises, and at others, they are fragments of the precarious urban fabric. This is the model of contradictory urbanization (re)produced between what is conceived and what is lived in the beaches of the metropolis.

In the case under analysis, the process of metropolization is not defined through the cooperation between municipalities. On the contrary, it is due to the production of spaces caused by overflows and the needs generated in the central nucleus or by the demands they attract. Coastal leisure has remained in the interstices and propitiated the (re)production of urban space, even when it is not included at the center of strategic actions.

State governments corroborate the promotion of this vector of metropolitan expansion, giving continuity to the implementation of urban infrastructure, acting as intermediaries in the articulations between local, regional and international companies, promoting the "vocations" of places, and granting environmental licenses for the construction of projects.

Establishing technical aspects and regulations regarding the use and occupation of the coast through enterprises that make *vilegiatura* viable is the burden of the municipalities and their (in)competencies. If, on the one hand, the possibility of filling the coffers through charging urban taxes (IPTU and ITBI) seems a significant advantage, on the other hand, the demands for public services outweigh and overburden the finances of municipal governments. There are apparent disparities in the metropolises of the Northeast between the hub city and the remaining cities.

The peri-urban coastal spaces have new spaces directly linked to the external demands of the host municipality. In this sense, the municipalities face the demands generated in and by the metropolis, with less integration and equality and stronger relationships of dependence and hierarchy among the municipalities subordinated to the metropolis. The (re)production of spatial fragmentation becomes a primary reality.

Even the technical-legal documents elaborated by the coastal municipalities are not implemented and observed as directed by legal precepts and duties of public administration. The most significant example is the Urban Development Directive Plan. The result of a descriptive technical diagnosis is as follows: the plan became the law which regulates and provides a foundation for the present and future of the city and the other urban spaces in the municipality. In most cases, it covers the urban coastal spaces and those suitable for the diffusion of modern maritime practices, especially by legalizing the construction of projects and the dissemination of large properties, based on low demographic densities. However, there are tremendous difficulties in conducting structural and corrective urban programs to correct the precarious conditions of the urban fabric. Municipalities lack the financial reserves and professional bodies to manage the Directive Plans. In truth, public executives search for and praise the arrival of tourist property projects without managing the fragmentation and the diversity of coastal locations.

References

Akyol D, Cigdem A (2016) Effects on the coastal areas of neoliberal urbanization in Turkey. Int J Agric Environ Res 02(06)

Barrantes-Reynolds MP (2011) The expansion of "real estate tourism" in coastal areas: its behaviour and implications. Recreat Soc Afr Asia Lat Am 2(1):51–70

Beaujour-Garnier J, Chabot G (1963) Traité de Géographie Urbaine. Armand Colin, Paris

Boyer M (2008) Les villegiatures du XVIe au XXIe siècle: panorame du tourisme sédentaire. Éditions ems, Paris

Brenner N (2014) Teses sobre urbanização. E-metropolis 19:6–26, ano 5

Burt JA (2014) The environmental costs of coastal urbanization in the Arabian Gulf. City 18(6):760–770. https://doi.org/10.1080/13604813.2014.962889

Corbin A (1989) O território do vazio. A praia e o imaginário ocidental. Companhia das Letras, São Paulo

Dantas E (2015). De ressignificação das cidades litorâneas à metropolização turística. In: Costa MCL, Pequeno R (eds) Fortaleza: transformações na ordem urbana. Letra Capital; Observatório das Metrópoles, Rio de Janeiro, pp 111–141

Debié F (1993) Une forme urbaine du premier age touristique: les promenades littorales. MappeMonde 1:32–37

Dedekorkut-Howes A, Bosman C (2015) The gold coast: Australia's playground? Cities 42:70–84

García-Ayllón S (2015) La Manga case study: consequences from short-term urban planning mass destiny of the Spanish Mediterranean coast. Cities 43:141–151

Hall CM, Müller DK (2004) Introduction: second homes, curse or blessing? Revisited. In: Tourism, mobility and second homes: between elite landscape and common ground. Channed View Publications, Clevedon (UK), pp 3–14

Instituto Brasileiro de Geografia e Estatística – IBGE (1980) Sinopse preliminar do censo demográfico 1980. Brasil, Rio de Janeiro

Instituto Brasileiro de Geografia e Estatística – IBGE (1980) (1991) Sinopse preliminar do censo demográfico 1991. Brasil, Rio de Janeiro

Instituto Brasileiro de Geografia e Estatística – IBGE (1980) (2000) Sinopse preliminar do censo demográfico 2000. Brasil, Rio de Janeiro

Instituto Brasileiro de Geografia e Estatística – IBGE (1980) (2010) Sinopse preliminar do censo demográfico 2010. Brasil, Rio de Janeiro

Latorre EM (1989) Genesis y formación de un espacio de ocio periurbano: Ribamontan al Mar (Cantabria). ERIA, pp 5–17

Lefebvre H (1999) A revolução urbana. Tradução de Sérgio Martins. EdUFMG, Belo Horizonte

Lencioni S (2013) Metropolização do espaço: processos e dinâmicas. In: Rua J et al (Orgs.). Metropolização do espaço: gestão territorial e relações urbano-rurais. Consequência, Rio de Janeiro, pp 17–34

Pereira AQ (2014) A urbanização vai à praia. Edições UFC, Fortaleza, Brasil

Persson I (2015) Second homes, legal framework and planning practice according to environmental sustainability in coastal areas: the Swedish setting. J Policy Res Tour Leis Events 7(1):48–61. https://doi.org/10.1080/19407963.2014.933228

Roca MN et al (2009) Expansão das segundas residências em Portugal. In: Anais do 1º Congresso de Desenvolvimento Regional de Cabo Verde. Cabo Verde, pp 2448–2474

Rutty M, Scott D (2014) Thermal range of coastal tourism resort microclimates. Tour Geogr Int J Tour Sp Place Environ. https://doi.org/10.1080/14616688.2014.932833

Terán F (1969) Ciudad y urbanización en el mundo actual. Editorial Blume, Madrid

Chapter 4
The Social and Urbanistic Effects of Tourism Developments in Fortaleza, Brazil

Abstract Nowadays, leisure activities, including summer vacations (through second residences) and tourism, have dynamized the occupation of the metropolitan coast in the Northeast of Brazil. The planning of these spaces pervades the public and private dimensions, and the logic of the production of space is defined by the emphasis on a technical-urbanistic approach. This article aims to understand the threads of this planning, focusing on the northeast metropolitan region of Fortaleza, in the State of Ceará. In order to do so, both quantitative and qualitative methodologies were used to obtain and process data and information: interviews, consultation of technical documents, and fieldwork. As a result, we concluded that the technical planning by entrepreneurs and the position of local governments point to the consolidation of spatial fragmentation in the coastal urban fabric. This situation is linked to the articulates with the promotion of urbanistic models, which value the use of private space to the detriment of public space.

Keywords Urban planning · Metropolitan areas · Maritime practices

The previous chapters examined the details of coastal urbanization on a regional scale. Without undermining the regional nature of our research, this chapter scrutinizes the characteristics, qualities, and processes in the intra-metropolitan scale, focusing on the metropolization of Fortaleza. We address the concept of geographic scale and assume that the processes in all the northeastern metropolises follow similar trends. Studies on the intra-metropolitan scale allow the use of qualitative methodological techniques and provide a level of detailed knowledge, which is not possible on the regional scale. Without a doubt, there is complementarity in the analyses and their respective results.

In summary, since the 1990s, the coast of the Northeast of Brazil has been recognized as a planned space for modern maritime practices (tourism and *vilegiatura*/second residences). The elements pointed out by Dantas (2013) indicate the relevance of these processes in contemporary times, which by changing the hierarchy of the urban network in question contributes to the restructuring of the role of the Brazilian coastal metropolises. As a result, the process of urbanization of the territory associated with the accumulation of leisure activities in the metropolitan coastal

© The Author(s), under exclusive license to Springer Nature Switzerland AG 2020   51
A. Queiroz Pereira, *Coastal Resorts and Urbanization in Northeast Brazil*,
SpringerBriefs in Latin American Studies,
https://doi.org/10.1007/978-3-030-46593-3_4

space is interconnected, principally in the three main metropolises of the Region (Salvador, Recife, and Fortaleza).

In other writings, we have prioritized the understanding of the process in the Northeast of Brazil, especially when analyzing the location of the quantifiable seasonal-use residences, the public policies of technification of the territory, the new resort-type projects, and the tourist-real estate complexes found on the metropolitan coastline (Pereira 2014). Many phenomena can be considered as metropolization vectors. These dynamics move between social, economic, political, and cultural dimensions, promoting interrelationships between the municipalities of the metropolitan region, cementing interdependencies (with different intensities and temporalities), especially with the hub city: the demand for leisure significantly promotes metropolitan spaces.

In the Northeast region, faced with the political-economic fragility of secondary urban centers, metropolitan spaces have emphatically become command spaces of the urban hierarchy (Regic 2007). The Metropolitan Region of Fortaleza (MRF) is a perfect example of this framework (Silva 2007). In demographic terms, the MR of Fortaleza concentrates 45% of the state's population (IBGE 2017). With a gross domestic product over R\$84 billion, the MRF has 60.9% of all the wealth produced in the state (IPECE 2016).

The location on the Atlantic coast is one of the most significant determinants of the metropolization of Fortaleza-Ceará-Brazil, as it is highly relevant to the effectuation of the metropolitan leisure space. Following current standards, the coastal environment has resulted in the development of spatially manifested modern maritime practices, through the process of metropolization as a function of leisure. In the 1970s and 1980s, *gatekeepers* played a role in the formation of the seaside resorts in metropolitan areas. From the 1990s, the internal production of these places was fragmented by the action of different real estate developers. There are varied dynamics, and the processes became more complex. Thus, this article discusses the diverse, and even contradictory, perceptions of the increased value of the coast. This process does not only involve real estate developers and vilegiaturists, or generally speaking, *vacanciers*. Here, we consider the unfolding of this phenomenon in more depth.

In this sense, it is relevant to consider the ideas—forces that mark the actions and discourse of the agents involved in the planning of metropolitan coastal spaces. Our objective is to present and evaluate how architects/town planners and professionals from the municipal public authority and residents of the coastal areas regard the consequences of the metropolization of maritime leisure in the metropolitan space most closely integrated to Fortaleza (the municipalities of Aquiraz, Caucaia, Cascavel, and São Gonçalo do Amarante). This methodological option relates technical-bureaucratic perspectives to everyday observations, based on popular knowledge, to understand the (re)production of urban space.

4.1 The Occupation of the Coast and the Critique of Urban Projects

The data mapped in Fig. 4.1 identify the amount and typology of private households registered by the official 2010 Census. In the four municipalities of the space in question, the focus is on the 17 coastal locations and the number of vacant, occupied, and occasional-use private households (considered second homes). Aquiraz and Caucaia have the most significant number, whereas, in Cumbuco and Porto das Dunas, there are enterprises with hundreds of units for sale. Occasional-use dwellings are an important determinant of the urbanization in these places, as they attract seasonal flows of vilegiaturists and exchanges, especially those related to the real estate market, given the significant number of vacant dwellings (most up for sale or rental). This situation results from cumulative processes lasting approximately 50 years. A kaleidoscopic coastal space has formed, marked by the diversity of socio-spatial uses and conflicts.

The metropolitan leisure space in Fortaleza has proportional impacts on the resident population of the municipalities involved (between 40 thousand and 300 thousand inhabitants). The role of leisure in the constitution of the real estate park and the urban forms present in the coastal area of these municipalities is worthy of consideration. It is a question of geographic, not cartographic, scale, where the intensity

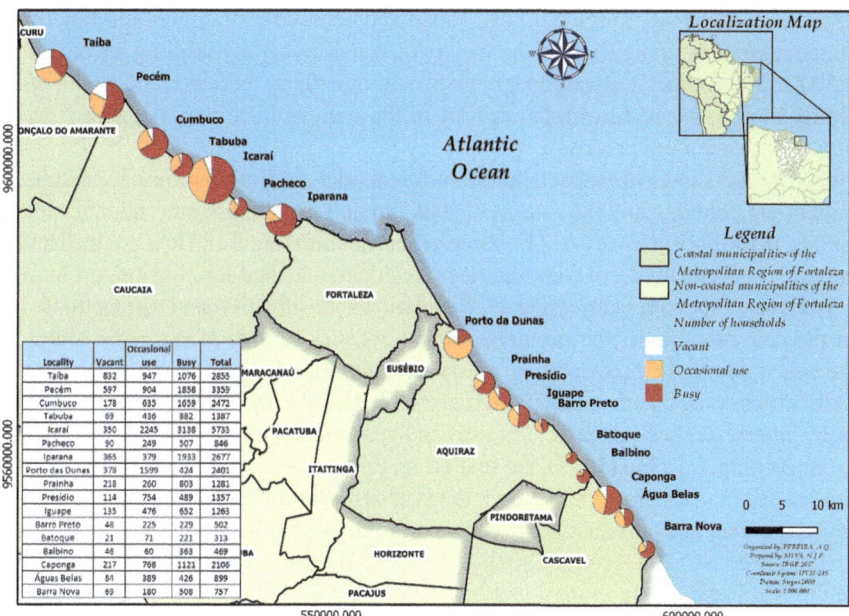

Fig. 4.1 Location of occupied, unoccupied, and vacant dwellings, in the coastal localities of the municipalities of Aquiraz, Cascavel, Caucaia, and São Gonçalo do Amarante, 2010. *Source* IBGE Census Synopsis (2010)

of the phenomenon is perceived in the face of the reality under study. Real estate for leisure (and/or mobilized for it) is the reason for the occupation of more than 50% of the territory in these localities, based on patterns of horizontal, linear, and discontinuous spatial incorporation.

The directions of the increased value of coastal spaces under the urban planning of Ceará's metropolis (northeastern Brazil) refer to what Lefevbre (1999) called the "urbanism of sales representatives." Planners in this category promote alleged, "places of happiness in a miraculously and wonderfully transformed daily life" (ibid., 25). Likewise, Harvey (2004) analyzes the production of spatial utopias, examining some examples (Disneyland, mega shopping malls, the suburbs), and relating them to what Louis Marin called degenerate utopias, "protected, safe, well-organized, easily accessible environments, above all, pleasant, relaxing and conflict-free" (ibid., p. 220). The metropolization of leisure understood as a process refers to the production of spatial and urbanistic forms with these characteristics and which have become intrinsic marks on the Northeastern metropolitan coastal space.

Due to environmental and territorial degradation in other coastal areas around the world destined to leisure (tourism and *vilegiatura*), parameters inhibiting coastal occupation are discussed and pointed to as necessary measures (Daligaux 2003; Cumbrera and Lara 2010).

The Integrated Coastal Management Project (Orla Project), proposed in 2002 by the Brazilian Federal Government, aimed to persuade coastal municipalities to draw up plans to mediate the use and occupation of their shorelines. The document demonstrates the seafront's political, social and economic strategies and functions. At the same time, it highlighted the variety of scenarios observed along the country's 8,500 km coastline. Leisure and tourism activities have become known for their ability to produce features/relationships in the various localities, integrating them into the urbanization process.

Figure 4.2 shows the most common urban models in Brazilian resorts, results of both urban planning and the actions of state/municipal governments and other social agents. In the peri-urban area of Fortaleza, there are horizontal and low verticalization patterns (A), with nuclei of communities of residents and land subdivisions promoting expansion spots with heterogeneous occupation (densification and rarefaction) (B). In the coastal capitals, the promenades and avenues to the sea give free access to the beach, but in the metropolitan coastal communities the pattern verified is that of indirect access (C), with narrow roads surrounded by a set of single-family residences or by enterprises with dozens of houses and apartments.

According to Debié (1993), the first urban forms linked to leisure and housing by the sea originated in Europe between the years 1850 and 1930. The promenades, the broad avenues, and other public spaces became models for the production of urbanism on marine coasts. However, in Brazil and, especially in the metropolization of leisure in the Northeast, the rearguard model and public access to the beach were eclipsed by the model of private appropriation.

In states such as Bahia, Ceará, Pernambuco, and Rio Grande do Norte, in the Northeast region, there is an accumulation of second homes in metropolitan spaces. Along the northeastern coast, traditional production patterns of second homes coexist

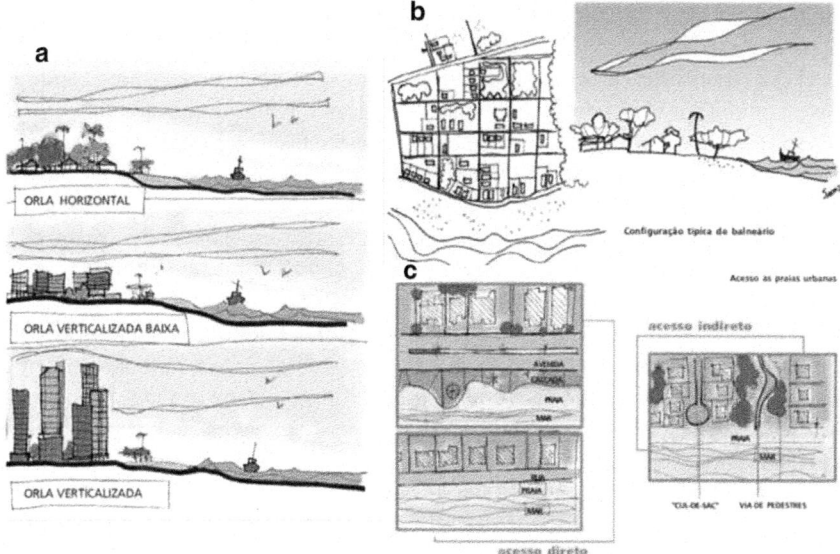

Fig. 4.2 Models and scenarios of the occupation of the Brazilian coast. **a** Models of the height of constructions on the seafront. **b** Predominant pattern of urban features in maritime resorts. **c** Urban features and models of access to the beach. *Source* ORLA PROJECT (2002)

with new models (tourism and real estate), which are based on complexes imitating globalized architecture and services (Algarve and Mediterranean). These urbanistic structures, common to coastal areas, are found in the environs of these states' main cities (Salvador, Recife, Natal, and Fortaleza). In addition to the flows of local visitors (owners of second homes), there are both national (other regions such as the Center-West and Southeast) and international visitors (mainly Portuguese, Italians, French, and Spanish).

In the evaluation of Fortaleza's city planners (names omitted for ethical reasons), there is no urbanistic concern for the state and the metropolis' coastline. The heterogeneity of the occupation is spontaneous and thus, lacks planning and urban design. For interviewee 2, in Ceará, the landscape potential of the coast has not been tapped. According to this interviewee, the recommendations found in the specialized literature indicate that successful interventions give an impression of zeal, cleanliness, water quality, social interaction, and enormous landscape opportunities.

For designers, the architectural dimension is individualized. In this regard, the fifth interviewee states that "although it is possible to identify relative quality (aesthetic, functional, technical-constructive) in projected buildings (summer homes, resorts, inns, hotels) by architects, this is not enough to affect the coastal public spaces."

The general evaluation is that there is no urbanistic and architectural formalism in the public coastal areas. The second interviewee considers that the common elements, sidewalks, squares, esplanades, and beach huts have patterns that do not fit the nearby urban context, which contributes to their consequent deterioration. The most democratic act would be to keep open areas "full of life and activities" (Interviewee 2).

These issues refer directly to the types of projects implemented (tourist and residential complexes, some calling themselves resorts), which converge in the discussion of the criteria formatted by planners, due to the taste and the demands of future buyers and users. These considerations show a model of urbanization of the territory close to degenerate utopias: they are enterprises "of great size, very luxurious, with all possible and imaginable services. Completely (sic) turned inwardly, with no relation to their surroundings" (Interviewee 1). Thus, they form "small holiday towns planned by private enterprises" (Interviewee 3).

Figure 2.3 shows a mosaic of images exemplifying the current occupation patterns and production of the coastal space in the metropolitan region of Fortaleza, more specifically in Porto das Dunas (Aquiraz-CE). The expansion of the urban network parallel to the coastline is evident, however, without any extension of public roads. The undertaking guarantees the owners of the apartments a closed and "safe" leisure environment, with access to the beach.

The planning and construction of this type of tourist-real estate complex is a new vector in the production of coastal space. Such ventures can be compared to "citadels of leisure" since they occupy areas of (on average) more than 200 hectares and offer a diversity of options for the seaside stay. In Ceará, two are at the commercialization stage with completed infrastructural works: *Aquiraz Riviera* and *Golf Ville*, both located in Aquiraz. Other ventures in the planning stage also have a preferential location in the metropolitan coastal space (in particular, Caucaia and Cascavel), thus demonstrating the primacy of relationships of metropolitan integration for projects of this size.

According to the professionals, the urban models and architectural forms of the enterprises (both complexes and condominiums) meet four basic requirements: comfort, homogeneous leisure practices, ostentation/social distinction, and, above all, safety. For interviewee 2, these requirements condition the production of real daydreams, which simultaneously exclude some issues that should be prioritized.

> I think an ideal project has an urbanistic part that preserves the visual aspect, coherent implantation taking into account the neighbors and the urban environment, sensitive to the sanitary question, the preservation, and increase of the native vegetation. The architecture should be simple, efficient, creative and without attention-seeking daydreams. (Interviewee 2)

Synthesized, these statements reveal multiple criticisms of the determinants of a problematic picture.

> The lack of planning and awareness-raising of the native population and even the politicians have transformed our coast into a large *favela*. There is a great lack of infrastructure (roads and airports) and aggregate services (schools and hospitals). Subdivisions occupy the areas indiscriminately. We need more quality in our coastal urban spaces if we are to attract tourism that brings income and improves the living standards of the local population (Interviewee 3).

> It is fundamental that architects have a critical conscience regarding the type of urban occupation and building in coastal spaces and their social and environmental impacts, so that

they can act with greater protagonism with the State, through planning, master plans, urban design, and environmental consultancies and the market, together with construction companies, developers and private clients. It is not an easy task, as it most often runs against the interests of real estate capital (Interviewee 5).

There is a consensus regarding the misconceptions about how urban space is produced on the coast of Ceará. The omission of the public authorities and the flexibility of the legislation were criticized, as well as the lack of inspection of what has been carried out. It is conceded that municipalities are unable to choose the urban forms suitable to the characteristics of the previous occupation of the territory, including those that harmonize with the natural landscapes. The planners criticize the role of architecture and urbanism, considering them as knowledge at the service of the interests of real estate developers.

On the other hand, the condemnations attributed to local residents are misplaced. The joining of political forces has left them weakened since any instruments of participation in the ordination of their location have been withdrawn. The orders come from the local, national, and international real estate and tourism dynamics that have promoted the coast according to the consumption of the spaces, governed by the urban-mercantile standard of access to urban land (Pereira and Dantas 2008).

The tourist complexes occupying hundreds of hectares are important expansion vectors for the coastal urbanization in the Northeast. However, their construction and operation pose challenges to territorial control and public administrations, especially the impacts on local communities (fishermen) and the possibility of environmental degradation. The circumstances evidenced in the Northeast are not an exception, on the contrary, they follow worldwide conditions (Akyol and Cigdem 2016; Burt 2014; Persson 2015; Barrantes-Reynolds 2011).

4.2   Local Public Power: Between Technique and Politics

Municipalities bear the consequences of the increased value of coastal space. Previously, territorial zoning and the patterns of use and occupation of spaces, as well as the definition of the urban areas conducive to urban expansion were the responsibility of local governments. Through omission or technical and political unpreparedness, the municipalities encourage or merely allow the occupation of coastal space, without a priori reflections on possible future demands.

The Urban Development Master Plan (PDDU) is a technical tool and basic policy (imposed by federal law) in the conduct of the occupation of municipal territories in Brazil. It sets out the consolidation zones of urbanization, also pointing to those that are a priority for public and private investments. In the case of the metropolitan coastline under analysis, all the municipalities elaborated their PDDUs with State Government resources. Table 4.1 describes the main features of diagnoses, zoning, and structural projects. Briefly, there is a clear alignment between current processes and what PDDUs prescribe and foster. They include the coastline (or the broader coastal space) in the metropolitan context as a *front* of urban expansion. This implies

Table 4.1 Characterization of the metropolitan coastal space according to the elements described in the Master Plans

Municipality	Diagnosis	Zoning	Structural projects
Aquiraz	Large real estate investments (Subdivisions, Resorts), culminating in large-scale real estate speculation Implantation of the Extractive Reserve of Batoque	Urban Expansion Zones—ZEU: II—Strategic Area of Coastal Interest—AEIL (Porto das Dunas, Prainha, Presidio, Iguape, Barro Preto and Batoque)	Structuring of the coastal strip of the Municipality Intervention in Critical Areas Requalification of the Highway linking Beach/urban core/Justiniano de Serpa
Caucaia	The attraction of Maritime Sports (Highlight for Cumbuco Beach) Installation of Large-Scale Resorts (Ex: Vila Galé Cumbuco)	Zona Urbana—Municipal Headquarters, Coastal (Zone of urbanization restricted by ecological fragility)	Works to recover the energy heat sink
Cascavel	Investments in the tourism sector on a small scale (lodges, beach huts)	Area 01: Limited to the north by the Atlantic Ocean, to the east by the Choró River, to the south by the CE-040 and to the west by the Mupeba Creek Area 02: Defined within its limits, it comprises the district of Cascavel; Area 03: It comprises the district seat of Caponga	Urbanization of coastal localities; Tidal containment works Implementation of new access infrastructure to Barra Nova beach
SG of the amarante	Implantation of the Pecém Industrial Port Complex. Leisure and tourism areas	Zone of Priority Urbanization (coastal strip of Taíba district)	Urbanization of the Coastal Range of the Taíba District Headquarters

Source Executive Plans of the municipalities of Aquiraz, Caucaia, Cascavel, and São Gonçalo do Amarante (2015)

the possibility of consolidating the processes mentioned in the previous topic, that is, the advance of the tourism-real estate sector.

If the Master Plans give an overall impression of the medium- and long-term actions of local authorities, we evaluated the need to verticalize the analysis to detect the public actions given the presence of real estate focused on leisure and tourism.

Second residences and other ventures related to maritime *vilegiatura* have a substantial impact on the production of the coastal space. The timespan of the process is of more than four decades of expansion, hence, the importance of identifying maritime *vilegiatura* (and other leisure practices) in the diagnoses and prognoses of local administrations.

Initial research confirmed that none of the municipalities maintain any registers or studies related to second residences or holiday homes (as they are known). There were two noteworthy reasons for this: (i) the consideration that second residences are not the responsibility of the municipalities, and (ii) the lack of strategic thinking on this issue. The first answer demonstrates the ignorance of municipal attributions in the government of their territory. The second can be explained by the precarious nature of the technical staff of the secretariats concerned. Of the four visited by the researchers, two did not have specialized professionals permanently included in the administration. The discussions involving the installation of large enterprises do not undergo technical evaluations in the areas mentioned above; instead, they are defined in higher instances and with eminently political and/or corporate endorsement. Given this technical and managerial fragility, the professionals interviewed admitted that on a scale of 1–5, where one is very important, and five is unimportant, second residences rank first in the organization of the coast of the respective municipalities.

Initially, the technical representatives were asked about the relationship between the spread of second home users and the development of tourist activities. In general terms, the respondents indicated that there is a proximity between the evolution of both activities. These considerations imply that the use and installation of second homes provide facilities for other tourists, provide a source of tourists, attract domestic tourism, add economic benefits to those already attracted by tourism, and are capable of instigating regional development.

To better understand the impacts caused by second residences and their users, a methodology developed by Müller et al. (2008) was used, in which the representatives of the municipal governments indicated the social, physical/environmental, and economic relevance of second residences. The data were subdivided into two groups: positive and negative (Table 4.2). The relevance was quantified on a five-level scale (1—very important to 5—unimportant). Four interviewees indicated by the tourism secretariats of the respective metropolitan coastal municipalities were interviewed.

In terms of positive social effects, local governments consider second homes as vectors of modernization that improve the living conditions of local residents. This is reflected mainly in the weight attributed to changes in the community's lifestyle and new ideas (customs, techniques, political ideologies, and beliefs, among others). In negative terms, the greatest emphasis was on the increase in crime, loss of cultural identity, empty properties during certain seasons, and changes in the social structure (forms of community and family organization).

Throughout the Brazilian Northeast (executive and legislative), municipal authorities face difficulties in preparing to manage the impacts derived from the circular migration motivated the use of second homes (Adamiak et al. 2017; Hall 2015). In reality, the gated condominium tourist spaces have privatized spaces and created their own regulations in order to enhance their real estate products and meet the

Table 4.2 Indexes of the relevance of the impacts of second homes in the municipalities of Aquiraz, Caucaia, Cascavel, and São Gonçalo do Amarante

Impacts		Fashion
Social effects		
Positives	Improvements in the community's lifestyle	1
	New ideas entering the community	1
	Creation of facilities	1
	Preservation of a traditional way of life	4
Negatives	Antagonism between owners of second homes and local residents	3
	Increases in crime	3
	Loss of cultural identity	1
	Empty properties during the low season	1
	Changes in traditional lifestyles	2
	Changes in the social structure	2
	Limitation of access to the beach and other recreational areas	3
Physical/environmental effects		
Positives	Beautification of the area	1
	Protection of natural areas	3
	Protection of historical heritage	1
	Protection of the traditional way of life	3
	The emergence of new services	2
Negatives	Loss of visual amenities	–
	Irregular waste disposal	1
	Environmental Degradation	1
	Overuse of the road system	1
	Deregulation of the traditional way of life	2
	Exploitation of natural areas	1
	Inadequate use of land in conservation areas	1
	Inadequate water supply	1
Economic effects		
Positives	Restoration of land value	1
	Increased employment opportunities	3
	Creation of a new economic base	3
	Creation of new industrial services	4
	Increase in duties and taxes	4
	Contribution to the maintenance of existing services	4
Negatives	Rising land prices higher than local gains	1
	Property price increase	1
	Increasing the cost of local goods and services	1
	Increased costs for local governments	1

Source Fieldwork, organized by the author (2015)

expectations of *vancanciers* and residents, which hinders the governance process of coastal spaces (Dredge 2015)

A more pessimistic view is detected in the production of physical-environmental effects. Addressing the positive and negative transformations, even without specific diagnoses, the technicians highlighted problems regarding solid waste disposal, the use of the road system, the use of water resources, and the degradation of vulnerable natural environments (beaches, mangrove swamps, and dunes). The notion of the beautification of the area was perceived as positive and refers to the standardization of constructions and the artificial, built environments, mainly in private big enterprises (large condominiums, *resorts,* and tourist complexes).

In economic terms, the positive factors are mainly in the real estate sector and job creation. The highest monetary values of urban land in the municipalities are in the coastal localities, especially land on the seafront. When mentioning job creation, the interviewees refer to the less formal services sector (with less need for technical knowledge), that is, the maintenance, surveillance, and cleaning services for real estate. In Ceará, generically, these domestic workers are known as caretakers. Other low-paid jobs (around a minimum wage, currently R$ 998.00) are provided by real estate and leisure ventures (chambermaids, waiters, security guards, janitors, gardeners, and kitchen helpers, among others). Besides these, there are also self-employed workers (plumbers, stonemasons, carpenters, and small business owners, among others).

Regarding the impacts generated by the collection of taxes and duties, municipal managers report the high rates of default on the part of seasonal owners of real estate. This is compounded by the assessment that the amounts collected are not sufficient to cover the demands generated (higher production of solid waste, greater demands on health services and prevention of epidemics, pressure for improvement of road infrastructure). These facts, perceived as problems, are added to the increase in property, goods, and services prices in the localities. These economic conditions mainly affect resident populations, who, on the whole, have low monthly family incomes.

Similarly, to the coastal cities in the Metropolitan Region of Fortaleza, field investigations demonstrate that the drinking water supply is a problem for coastal urbanization in the Northeast. Second homes and large-scale real estate and tourism developments use groundwater from deep wells. Thus, there is a danger of emptying or salinizing the aquifers. The data of the Environmental Impact Reports also demonstrate the absence of an efficient sewage network. In general, mega-enterprises are required by law to build individual waste treatment stations. However, second homes, especially the older ones, use septic tanks, a technique that is unable to contain the contaminating effects of organic waste. This leads to an expansion of the problem, due to the possibility of underground water resources, which are the main supply source in these urbanized areas, being contaminated. These characteristics demonstrate that the urbanization process is incomplete and has a precarious basic infrastructure.

The states seek to improve the occupation of coastal spaces through studies such as Economic-Ecological Zoning, which raise the concern with sustainability and the conservation of natural environments (and their functions). However, it is also similar to the acceleration of erosion processes on the Atlantic coast due to occupations in inappropriate areas acting as a barrier to the natural flow of matter and energy.

It is important to remember that since the end of the twentieth century, these municipalities have drawn up and approved the Urban Development Master Plans (PDDU). In theory, this means they should produce a diagnosis of the processes that drive urbanization in the municipal territory and carry out an arsenal of studies, to act as the foundation of a range of regulations (in particular, the Law on Use and Occupation and the Code of Works and Positions), democratically approved by the municipal councils. It can be inferred that these studies and laws would increase the degree of municipal autonomy, also enabling them to resolve undesirable scenarios and conditions affecting the majority of the local population. In the specific case under analysis, faced with these transformation vectors the municipalities are weak, and there is no short-term or medium-term perspective of measures for participatory discussions on what is or is not of interest to the townspeople.

4.3   The Residents of the Beach and Their Conceptions

The conceptions by which the groups and entities define processes/activities are decisive in the understanding of the socio-spatial transformations. Having addressed the perceptions of urban municipal professionals, this section focuses on the residents' view. How do they interpret the activities that predominate and define the uses and investments in the spaces where they live? What temporality governs the concepts present in their discourse? Is there a single resident profile, a single thought? The answers to these questions are discussed below.

One of the best-remembered themes when discussing tropical coastal spaces is precisely the relationship between the demands of the *vacanciers* (tourists, day-trippers, vilegiaturists) and the social organization of the inhabitants of these places. Undoubtedly, there is a range of criteria used to differentiate these social subjects; however, in the geographical approach to the problem, relationships with the same place contribute significantly to the elaboration of this distinction. For Cazelais (2009), the locals are the parameter. They hold empirical knowledge of the place and a strong relationship of intimacy (self-recognition). When identifying themselves, the place of residence is one of the first defining elements of their social existence. On the other hand, the discussions about modern nomadism and the formation of multi-territorialities, without being an absolute condition, remind us that the condition of the inhabitant is also open to transience.

According to Cazelais (2009), the residents have "*la propriété morale*" (a feeling of identification) and "*la propriété foncièr*" (Based on legal and financial instruments). If there is a certain consensus regarding the former, this does not apply to the latter, given the growing worldwide movement of residential real estate acquisition due to tourism and *vilegiatura*. In space-time, the definition of resident and *vacancier* is relational. The author shares this view by stating that "*des villégiateurs de longue*

Table 4.3 The coastal space under analysis according to the municipality, population, and territorial area

Municipality	Population	Area (km^2)	Demographic density (hab/km^2)	Extension of coastline occupied by leisure (km)
Aquiraz	72,628	482.38	150.50	17.5
Caucaia	325,441	1,228.51	264.91	~25.3
Cascavel	66,142	835.00	78.79	8.10
São Gonçalo do Amarante	43,890	834.45	52.60	~16.7

Source IBGE (2010)

date serant quand même consider comme des'étrangers' par les résidents, molgré leurs efforts parfais répétes et soutenus pours s'intégrer" (IBIDEM, p.182).

Lefevbre's (1999) thinking about daily life offers a coherent differentiation between the two social subjects under evaluation. The *vacanciers* consider leisure space-time as a way to escape from everyday life, viewing it as complementary to their reproductive relations. For the residents, the same space has a notion of differentiated time: a notion of ordinary time, without surprises, proper to daily obligations (work, subsistence).

With the exception of Caucaia, the municipalities have a total population below 75 thousand inhabitants, with a territorial area of more than 400 km^2. This results in a low population density index, especially if compared with Fortaleza (7,786.44 inhab./km, according to IBGE in 2010). The distribution of the urbanization of the territory follows two primordial locations: the district seat of the municipal administration (the city) and the spaces bordering the Atlantic Ocean (the coast). The coast is the base location of both traditional communities (fishers, artisans, and others) and the second homes, hotels, and tourist-real estate complexes. Thus, leisure-related urbanization is a decisive factor in understanding this metropolitan area; it occupies thousands of hectares along tens of kilometers of coastline (Table 4.3).

In contemporary times and the coastal locations of the Fortaleza Metropolitan region, and perhaps the others in the Northeast, the figure of the resident cannot be restricted to the traditional groups (fishers and shellfish fishers, among others). It is increasingly common for "recent" residents to be attracted by the process of the increased value of coastal areas and the resulting socio-economic transformations. Starting from this understanding, during the fieldwork, residents with different profiles interviewed (from those who self-identified strongly as *natives* to the recently installed). According to data from the 2010 demographic census, there is a spectrum of cases: locations with 850 inhabitants to small neighborhoods with more than 10,000 residents (Fig. 4.3). Relatively low demographic figures coexist with real estate occupied by thousands more during high season. The case of Porto das Dunas in the municipality of Aquiraz is emblematic. The 1200 people registered in the census are multiplied dozens of times in periods of vacation or even national holidays.

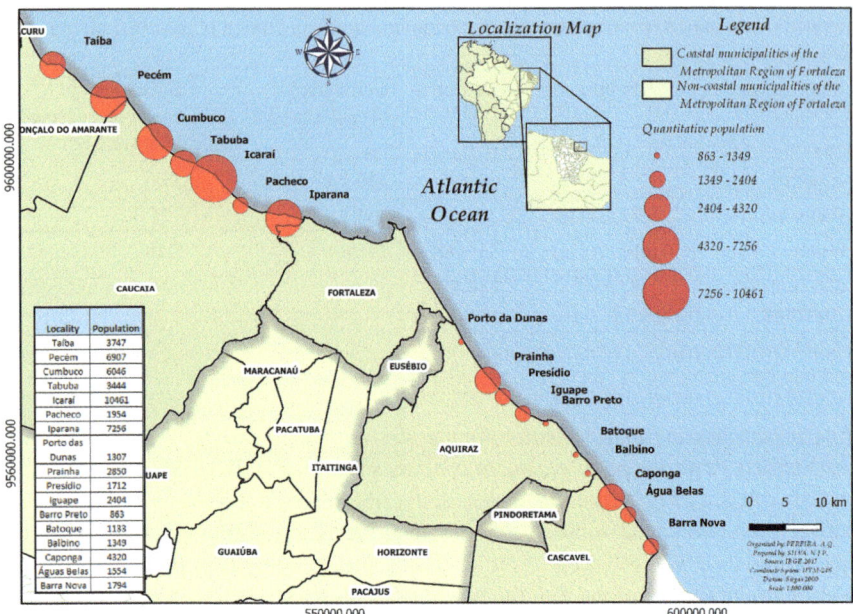

Fig. 4.3 Resident population, according to coastal localities of the municipalities of Aquiraz, Cascavel, Caucaia, and São Gonçalo do Amarante, 2010. *Source* Synopsis of the IBGE Census (2010)

Absolute numbers indicate the production of these spaces; however, the conceptions regarding the presence of *vacanciers* are an important aspect of this study. During the interviews, three profiles of residents were verified: i. traditional inhabitants of locations incorporated by the dynamics of *vilegiatura* and tourism; ii. residents of localities with social movements who resist the socio-economic developments of modern maritimity (mainly, the commodification of land); and iii. recent inhabitants attracted precisely by the transformations of the last decades.

The first group is mainly made up of people who have followed the transformations that have occurred during the last 40/50 years. When the older people speak, there is resentment regarding the "lost land." Many feel cheated by promises (never fulfilled, according to them) made by the gatekeeper developers. The residents view the conditions of the past with the experienced gaze of the present. The land of before, the dwelling place, where fences did not exist, is now remembered as an abstract space, geometrized in lots, which is measured in reais, euros, or dollars. They live daily with the discomfort of the inheritance that no longer exists. Younger residents, armed with stories told by their parents and grandparents, admit the naivety of their ancestors, justifying that at the time they could never have imagined what was to come.

The residents look at the place today and share their memories. They refer to the straw houses, which were dotted randomly, on the sands. The land by the sea was not for habitation; there were the huts that sheltered the small rafts, where fishermen talked to each other. Speaking of "the outside world," they mention their rare comings and goings because access and means of transportation were non-existent or very precarious. The sun and the kerosene lamp were the only sources of light.

In view of the conditions outlined above, together with the almost complete absence of free public services in days gone by, residents praise the newer basic infrastructure, especially roads, means of transport and communication, and electrification. However, they admit that most of the "improvements now" are due to other activities and not a direct response to their needs.

Other advances pointed out by the residents refer to the economic dimension. Wage labor and a monthly income are cited as gains, mainly due to the possibility of increased consumption. This is not surprising as the residents themselves say that it is increasingly difficult to "get sustenance" (biological and social reproduction) in the primary and artisanal activities of the past (fishing, shellfish fishing, small plantations, and small animal raising). Young people criticize the socio-occupational alternatives; they have few choices other than the professions of caretaker (general services) and waiter.

In what was said and unsaid, it is evident that the residents perceive themselves as coadjutants in the processes that now define their lives. For them, the present conditions are a product of exterior forces. They do not think they have the potential to define what they need or achieve it. Empirically, they recognize the social problems that currently afflict them (fear of potential criminality, prostitution, and increased cost of living), they perceive them as by-products of the socio-spatial changes, but instantly return to the condition of beings led by a predefined destiny.

In the places that promote a model of local organization that resists the commercialization of land, the discourse is not homogeneous. There are conflicts of interest among the residents. The main reason for this is that although the division of land ownership is a collective decision, the infrastructural conditions in these localities are inferior to those in other places. For the residents opposed to the current situation, if the beach were open to various private leisure enterprises, the local economy would be more dynamic. The model of community organization is recent, and according to community leaders there are no real partnerships with local and state public authorities, they state that the majority of the population is in favor of blocking land negotiations.

The difficulty of organizing alternative sources of solidarity economy in the communities is another obstacle to higher levels of popular autonomy. In order to generate greater social cohesion about the elaboration of models of social organization, an interesting proposal would be the organization of municipal forums with the different communities and residents living in the beach areas. The joint exposition of their problems and the solutions found would mediate future decision-making, clarifying what actions are appropriate or not for each case.

Unlike the other residents, the views of the more recent arrivals, many of them attracted by the increase in tourist activities and *vilegiatura,* are based solely on the present situation. This type of resident defends the continuous "modernization" of the locality, that is, total opening to the process of implantation of tourist and real estate enterprises. Furthermore, they believe that this path that produces indirect gains to the residents is more effective than actions that oppose the increase in the number of visitors and vilegiaturists. They view the resulting problems as similar to other places and thus tolerable. They also believe that the presence of the "outsiders" is a source of help and attracts many benefits.

The residents' reports make it possible to describe the *vacanciers*, especially the vilegiaturists. Initially, there are the pioneers, the metropolitan vilegiaturists. In the early years of expansion, in the communities where the *villas* were built the vilegiaturists became known as "doctors" and "bosses." Not all of them were doctors or lawyers, but the local people used these forms of address to show their respect to the new arrivals, who had more schooling, influence, and possessions. Over time, they became recognizable by the population and interacted either by hiring services (hence the term boss) or by providing favors. There are countless reports: doctors who attended in their homes and donated medicines; others contributed financial resources to patronal parties, facilitated access to services in the Capital, and provided employment. The townspeople recognize that the vilegiaturists who settled 20 or 30 years ago were more ostentatious. According to the residents, they were looking for tranquility and rest during their stay.

In a contemporary view of the process, the locals admit the popularization of the visitors, and with a tone of prejudice, they affirm to the increasing arrival of visitors considered *pão-com-mortadela (Bread with bologna)*. This describes visitors who do not care about the quality of their accommodation, seek excitement, sun, and fun, spending as little as possible on their stay. They purchase smaller second homes with no architectural refinement.

Concerning the real estate dynamics, local merchants affirm that the previous model, where the self-build of individual houses predominated, was more beneficial to the community's economy, both through the sale of products and the hiring of local labor. It is worth mentioning that this pattern still exists to a lesser extent; however, it has been quickly replaced by the construction business (builders and developers) of condominiums.

If the autochthon vilegiaturists have contributed for decades to the redefinition of the daily lives of the residents, the participation of foreigners in this fabric has increased in the last decade. As already mentioned, the fieldwork confirmed the presence of foreigners (both vilegiaturists and residents) in the vast majority of the communities visited.

In their testimonies, residents distinguish between local vilegiaturists and foreigners in several ways. First, *gringos* are classified as tourists. Asked about the criteria used, the locals refer to distance and nationality: tourists come from far away, their "speech" is strange, it is exotic, and it is different. Language is one of the main impediments to social relationships. The investigations indicate that very few local residents speak languages other than their mother tongue. In part, this may explain the depiction of the allochthon vilegiaturists as reserved and little interested in establishing social contacts with the locals. Vilegiaturists who do not plan for greater proximity with the local populations acquire (or rent) their properties in condominium-type developments and resorts, where there are already owners (or tenants) of the same nationality. Examples of this nuance are the French in Aguas Belas, and with Scandinavians in Taiba, Cumbuco, and also in Porto das Dunas.

However, there are exceptions to this general case. Residents of communities such as Taíba, Iguape, Caponga, and Colônia describe those considered as "pioneers." These are foreigners who arrived first (and continued to attract others), before the new condominiums and resorts. They established business relationships with allochthon vilegiaturists or local landowners, acquired houses and land, and became known by the community. Some learnt the Portuguese language and/or taught their native language. Going beyond the basic forms of socializing and sociability, there are many instances of legal love matches between foreign vilegiaturists and residents. It is not possible here to quantify the intensity of this phenomenon, but one of the consequences of this situation has been the permanent migration of foreigners and also of local people.

Residents acknowledge that foreigners are becoming vilegiaturists (mainly because they own property) and the resulting more extended periods of stay (from weeks to months) have added new demands for leisure to the social dynamics of localities during different seasons practiced by vilegiaturists and local tourists. If the localities become internationalized by investments, visitors, and different demands, time in these communities is restructured according to the diversity of social subjects.

Unequivocally, the case of the Metropolitan Region of Fortaleza resembles other coastal locations in the Northeast, representing the Global South. Market planners, real estate developers, and politicians view tourism as a modernizing activity to activate the regional economy (new real estate park, marketing the location, and new jobs). This narrative considers the positive effects of the process.

Problems in the disarticulation between tourist uses of the territory and other uses have been reported in several countries, but with different approaches (Hall 2014, 2015; Adamiak et al. 2017). As shown by interviews with planners and residents (especially the poorest), there is strong evidence of social problems resulting from the touristification process: underemployment and seasonality, social-residential segregation from the formation of precarious housing, restriction of public uses along part of the seafront, and deregulation of traditional economic activities (fishing, agriculture, and animal extraction).

The choice of ending this text interacting with the discourse and the apprehensions of the residents was not unconsidered. Undeniably, the new and old patterns of consumption of space due to the *vacanciers'* desires accelerate the process of creating the metropolitan network of available coastal sites. In some places, this means economic, social, and cultural changes. However, there is a warning: such transformations will be democratic when scientific discourse and technocratic planning truly listen and consider as paramount the analyses, desires, and objections of different groups of residents.

References

Adamiak C, Pitkaïnen K, Lehtonen O (2017) Seasonal residence and counterurbanization: the role of second homes in population redistribution in Finland. GeoJournal 82:1035–1050

Akyol D, Cigdem A (2016) Effects on the coastal areas of neolieral urbanization in Turkey. Int J Agric Environ Res 02(06)

Barrantes-Reynolds, M P (2011) The expansion of "real estate tourism" in coastal areas: its behaviour and implications. Recreation and society in Africa Asia and Latin America 2(1):51–70

Burt JA (2014) The environmental costs of coastal urbanization in the Arabian Gulf. City 18(6):760–770. https://doi.org/10.1080/13604813.2014.962889

Cazelais N (2009) L'espace touristique: relations entre résidents, visiteurs et paysages. Ateliê Geográfico 3(1):179–193

Cumbrera MG, Lara EL (2010) Consecuencias del turismo de masas en el litoral de Andalucía (España). Caderno Virtual de Turismo 10(1):125–135

Daligaux J (2003) Urbanisation et environnement sur les littoraux: une analyse spatiale. Rives méditerranéennes 15:1–8

Dantas EWC (2013) Touristic metropolization in an industrialized region by monoculture. Rev Mercat 12(2). Recuperado 2015-04-06, de. http://www.mercator.ufc.br/index.php/mercator/article/view/1175

Dredge D (2015) Coastal and marine tourism. Exploring the prospects for emerging ocean industries to 2030. In: OECD workshop, Gothenburg, Sweden, June 2015

Hall CM (2014) Second home tourism: an international review. Tour Rev Int 18(3). https://doi.org/10.3727/154427214x14101901317039

Hall CM (2015) Second homes planning, policy and governance. J Policy Res Tour Leis Events 7(1):1–14. https://doi.org/10.1080/19407963.2014.964251

Harvey D (2004) Espaços de Esperança. Loyola, São Paulo, Brasil

REGIC. Regiões de influências das cidades (2007) Instituto Brasileiro de Geografia e Estatística – IBGE. Brasil, Rio de Janeiro

Instituto Brasileiro de Geografia e Estatística – IBGE (2010) Sinopse preliminar do censo demográfico 2010. Brasil, Rio de Janeiro

Instituto de Pesquisa e Estratégia Econômica do Ceará – IPECE (2016) *Perfil Básico Regional 2016* Região metropolitana de Fortaleza. Fortaleza, Brasil

Instituto Brasileiro de Geografia e Estatística – IBGE (2017) Estimativas da população. Brasil, Rio de Janeiro

Lefevbre H (1999) A revolução urbana. Tradução de Sérgio Martins. Belo Horizonte: Ed. UFMG

Ministério do Meio Ambiente (2002) *Projeto Orla*: fundamentos para gestão integrada. Brasília, Brasil

Müller DK, Hall CM, Keen D (2008) Second home tourism impact, planning and management. In: Hall CM, Müller DK (ed) Tourism, mobility and second homes: between elite landscape and common ground. Channed View Publications, Clevedon, UK, pp 15–34

Pereira AQ (2014) A urbanização vai à praia. Edições UFC, Fortaleza, Brasil

Pereira AQ, Dantas EWC (2008) Veraneio marítimo na metrópole: o caso de Aquiraz-CE. Sociedade & Natureza 20(2):93–106

Persson I (2015) Second homes, legal framework and planning practice according to environmental sustainability in coastal areas: the Swedish setting. J Policy Res Tour Leis Events 7(1):48–61. https://doi.org/10.1080/19407963.2014.933228

Silva JB (2007) Região metropolitana de Fortaleza. In: Silva JB et al (eds) Ceará: um novo olhar geográfico. Edições Demócrito Rocha, Fortaleza, pp 101–124

Conclusion

It is not difficult to perceive that there are spatial-temporal distinctions between occupations such as the French Riviera, the Cancún complex, and the metropolitan coast of the Brazilian Northeast. The times and wealth materialized in space have produced various urban densities. However, there is a common desire for leisure nascent in globalized urban society.

The coastal area of northeastern Brazil is renowned for its sandy beaches, sunny days, and the warm waters of the Atlantic. On the coastal plain, there are highways, condominiums, tourist-real estate, and resorts. Thus, the area attracts thousands of tourists and vilegiaturists who demand the local advantages associated with an urbanized space with the infrastructure of the city agglomerations.

Maritime *vilegiatura*, second homes, and tourism are practices and contents produced and disseminated as a by-product of the urban, capable of generating urban forms different from those we call cities. Especially in the Portuguese and Spanish literature, it is clear that the use of the concept of the second residence is imprecise. This cannot essentially be classified by the quantitative dimension, but rather by the qualitative one. Socially and geographically, not every property used for vacation or tourism can be considered a second home. But every *vacancier* who turns a property into a second home for themselves develops the practice of *vilegiatura*. The new online real estate sharing platforms make this relationship even more flexible.

In Brazil, until the first half of the twentieth century, the beach-city-leisure relationship was primarily internalized in the urban fabric of large cities. From the last two decades of the twentieth century, and especially at the beginning of the twenty-first century, the spaces produced for temporary stay (leisure and rest) have fostered the expansion of the metropolitan urban fabric. The integration between spaces, characteristic of metropolises, is accentuated, given the production of the network of territories for leisure. The metropolis is a decisive element in the consolidation of all flows and permanences. Besides the hub city, the metropolis is formed of subspaces with specific and integrated features and functions. Thus, metropolitanization in the coastal space, motivated by *vilegiatura*—tourism complements and

© The Author(s), under exclusive license to Springer Nature Switzerland AG 2020   69
A. Queiroz Pereira, *Coastal Resorts and Urbanization in Northeast Brazil*,
SpringerBriefs in Latin American Studies,
https://doi.org/10.1007/978-3-030-46593-3

updates the territorial division of labor and consumption in the internal structure of the urban-metropolitan fabric in the Northeast. Spatialities emerge as a function of leisure.

On the local scale, there are differences in the nodes of the leisure network, the places, and the effects of maritime *vilegiatura*. In many municipalities in the Northeast, the consolidated urban fabric on the coast is the most significant expression of urbanization in the respective spatial area. Real estate for seasonal use, economic activities, and even temporalities in these places are governed by the dictates of the metropolis(es). It is the time of the metropolis, free time, holidays, and vacations (non-work), which are responsible for the cadence of the transformations in seaside spaces governed by this rationale. The metropolis mediates and attracts flows from other metropolises on a national and international scale. It is the point of departure and arrival, responsible for the distribution of flows.

The planning of tourist territories is carried out by economic and political agents. For the former (the companies that organize the itineraries, conduct the flows, manage the accommodation, and build the real estate for *vacanciers*) generally, planning involves a specific scale of the activity and enterprise itself. Territory is a means to an end. Such corporations establish their interests with local governments, guarantee tax advantages (tax reductions), and different environmental conditions for their properties. When the business does not meet their expectations, the master plans are quickly adjusted. The market creates demands and trends/tastes (new services, real estate-tourist hybrids, festivals, marketing, construction of attractions) that are as diverse as possible.

For the political dimension, planning is a mechanism for structuring space or urbanizing it better. The State, at the federal and state levels, provides enormous financial resources for the technification of space and tourism. On the municipal scale, there are problems such as land use regulation and urban planning, the provision of collective services, which are aggravated by seasonality (which generates periods of intense demand). The inhabitants of these metropolitan spaces, in their multiplicity of profiles, incorporate, work, criticize, live, isolate themselves and interact with the vilegiaturists, and increase in volume. Vilegiaturists and tourists, in turn, prefer leisure, sports, and/or rest.

In the city, the search for nature and the non-city, usually described as an incentive for the stay on the beaches, reproduces the urban way of life in these localities (agglomeration, simultaneity, contradictions) and, thus, the new users reconstruct the image of paradise, residents, and investors, who, paradoxically, turn it into an urbanized space.

In terms of planning, public administrations in the Northeast, and their technocrats, when considering the objectives of the private sector (tourism and real estate) should rethink the production of coastal space, taking into account both the increasing urbanization and the variety of uses and users. They must demand larger counterparts from the entrepreneurs. The development of maritime *vilegiatura* (and other modern maritime practices) cannot occur to the detriment of the interests of the residents of these spaces, even because real estate developers are planning a hybrid coastal area, marked by tourism, by *vilegiatura*, and also by permanent residence.

With the constant allocation of public resources in infrastructure in the metropolitan area and the marketing of real estate launches, it is prudent to say that the *vilegiatura* tourism sector will tend to remain concentrated on the coast of the northeastern metropolises, especially in the municipalities where they have already built tourist complexes in Salvador, Recife, and Fortaleza. Tourist-real estate is a growing reality, even in the face of global economic crises, producing new ventures in the short and medium terms in these states.

The technical planning of the entrepreneurs and the positioning of local governments point to the consolidation of the spatial fragmentation of the coastal urban fabric. This is articulated by the promotion of urban models that value the use of private space to the detriment of the public space. Considering the precarious infrastructure in the spaces outside the tourist enterprises (basic sanitation, pavement of roads, and public health services), this is a major problem.

There are serious problems in relation to the planning carried out. It is necessary to emphasize that, even in coastal areas in the process of metropolitanization and that receive a significant number of tourists, the ineffectiveness or inexistence of systems for collecting/treating waste and supplying water supply is lasting and worrying condition. The concern stems from two situations: i. the rise in the artificialization of the territory with underground water extraction, without a collection of sewage and treatment network and ii. the claims of municipal administrations (local authorities) that they do not have the public resources capable of meeting the demands for drinking water and waste treatment. If the current model adopted continues, the result will be precarious tourist urbanization.

The participation of the Northeastern coastal residents (traditional or non-traditional communities) in the elaboration of legislation is still timid, as is the allocation of public resources. The quality of the natural environment is a critical condition for populations and to maintain this process, despite these polluting activities are increasingly observed in the beach areas. Local authorities must pay attention to this vital issue. After all, dealing with environmental problems is an action that demands a vast volume of economic resources.

The mitigating measures presented in the plans for real estate-tourism ventures do not address the problems arising from their installation. However, it is worth mentioning that there are examples of small communities organized against the model of economic activity intended for the coast. Some have even been able to protect their lands through federal decrees and the creation of conservation units for sustainable use.

Tourist-real estate ventures and second homes are clear and, at times, preponderant expressions of urbanization in sections of the Brazilian coastal zone, including in the metropolises. Due to their importance and the problems presented here, it is clear that a single model does not meet the socio-economic needs of the Northeastern metropolitan coastline. In addition to taxes levied on large enterprises (resorts, theme parks, tourist complexes, condominiums with dozens of single-family units), municipalities should draft rules and laws, based on the City Statute, in order to request that before the installation of projects, the business groups responsible should provide infrastructural counterparts in the location where these ventures will be installed.

Focusing solely on the urban and architectural quality of the enterprise itself tends to degrade the urban and social quality of the overall coastal space that results. In terms of management and planning, in the medium and long terms, the problems will become insoluble and costly to the public coffers.

Index

Printed by Printforce, the Netherlands